Synthese Library

Studies in Epistemology, Logic, Methodology, and Philosophy of Science

Volume 416

Editor-in-Chief

Otávio Bueno, Department of Philosophy, University of Miami, USA

Editors

Berit Brogaard, University of Miami, USA
Anjan Chakravartty, University of Notre Dame, USA
Steven French, University of Leeds, UK
Catarina Dutilh Novaes, VU Amsterdam, The Netherlands

The aim of *Synthese Library* is to provide a forum for the best current work in the methodology and philosophy of science and in epistemology. A wide variety of different approaches have traditionally been represented in the Library, and every effort is made to maintain this variety, not for its own sake, but because we believe that there are many fruitful and illuminating approaches to the philosophy of science and related disciplines.

Special attention is paid to methodological studies which illustrate the interplay of empirical and philosophical viewpoints and to contributions to the formal (logical, set-theoretical, mathematical, information-theoretical, decision-theoretical, etc.) methodology of empirical sciences. Likewise, the applications of logical methods to epistemology as well as philosophically and methodologically relevant studies in logic are strongly encouraged. The emphasis on logic will be tempered by interest in the psychological, historical, and sociological aspects of science.

Besides monographs *Synthese Library* publishes thematically unified anthologies and edited volumes with a well-defined topical focus inside the aim and scope of the book series. The contributions in the volumes are expected to be focused and structurally organized in accordance with the central theme(s), and should be tied together by an extensive editorial introduction or set of introductions if the volume is divided into parts. An extensive bibliography and index are mandatory.

More information about this series at http://www.springer.com/series/6607

Ana-Maria Crețu • Michela Massimi
Editors

Knowledge from a Human Point of View

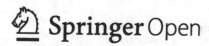
Springer Open

Editors
Ana-Maria Crețu
School of Philosophy, Psychology and
Language Science
University of Edinburgh
Edinburgh, UK

Michela Massimi
School of Philosophy, Psychology and
Language Science
University of Edinburgh
Edinburgh, UK

Synthese Library
ISBN 978-3-030-27043-8   ISBN 978-3-030-27041-4   (eBook)
https://doi.org/10.1007/978-3-030-27041-4

This Springer imprint is published by the registered company Springer Nature Switzerland AG.
The registered company address is: Gewerbestrasse 11, 6330 Cham, Switzerland

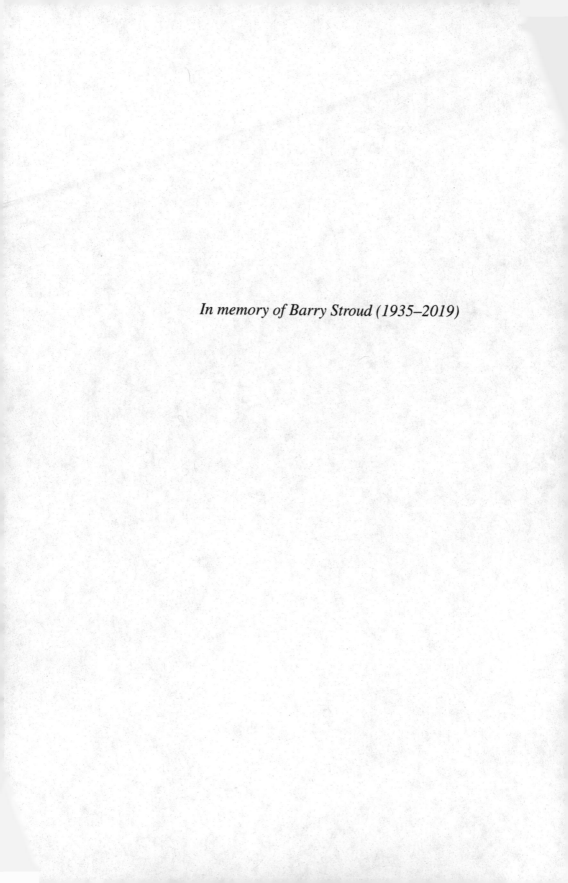

In memory of Barry Stroud (1935–2019)

Acknowledgements

The editors, Ana-Maria Crețu and Michela Massimi, are grateful to all the authors who contributed to this volume for their engagement with the topic and the many stimulating conversations during the conference *Knowledge from a Human Point of View*. Our thanks also go to the Editor for the Springer Synthese Library Series, Otávio Bueno, for enthusiastically supporting this project from the beginning. This edited collection is the research output of a project that has received funding from the European Research Council (ERC) under the European Union's Horizon 2020 research and innovation program (grant agreement European Consolidator Grant H2020-ERC-2014-CoG 647272, *Perspectival Realism. Science, Knowledge, and Truth from a Human Vantage Point*). We are very grateful to the ERC for supporting our research in this area.

Introduction

This edited collection of nine original essays was commissioned as part of the ERC Consolidator Grant project *Perspectival realism. Science, Knowledge, and Truth from a Human Vantage Point*. The guiding idea behind it is to explore the view known as "perspectivism" in philosophy of science by looking at its broader historical and epistemological context. Perspectivism in philosophy of science is often presented as a view about our scientific knowledge being historically and culturally situated. The scientific knowledge we can afford is inevitably the outcome of modelling practices, scientific theories, experimental techniques, conceptual resources inherent in specific 'scientific perspectives' that we—as historically situated epistemic agents—happen to occupy. Therefore, it is common currency to refer to the 'Newtonian perspective', or the 'Maxwellian perspective' (among innumerable others across the sciences) as a way of marking and specifying the particular *vantage point from which* knowledge claims are typically made. But what is philosophically at stake in this seemingly platitudinous move remains to be clarified. For one, if our scientific knowledge is indeed historically and culturally situated, can it ever be knowledge of the world *as is* (as opposed to knowledge of the world *as seen through our perspectival lenses*)? Relatedly, how does perspectivism affect the very notion of knowledge (qua justified true belief, under the traditional view) if justification and truth are themselves couched as perspectival notions?

This edited collection locates perspectivism within the wider landscape of history of Western philosophy and current epistemology. Two overarching questions guide the inquiry in the following chapters. When did the idea of knowledge from a human point of view emerge in the history of philosophy? And what role does the idea play in contemporary debates in epistemology? Each question invites more than one answer and the selection of chapters that follow is intended to give a brief—almost pointillistic, but nevertheless illuminating—introduction, rather than a comprehensive and exhaustive treatment of the topic. In what follows, we briefly introduce each chapter and the underlying narrative and leitmotiv that connects the first part of the book (with more historical analyses) to the second part (dedicated to ramifications in contemporary epistemology).

Situating perspectivism in the history of Western philosophy means locating a

distinctive notion of 'knowledge from a human point of view' as an emerging influential trend with far-reaching ramifications in contemporary epistemology.[1] When
did the epistemic agent's point of view become relevant in philosophical discussions about knowledge? The question might sound *prima facie* trivial (of course,
knowledge is necessarily from a human point of view—whose else's point of view
could it be?). But, in fact, it conceals a more profound issue. It has become a platitude (almost a cliché) to identify Kant in the history of Western philosophy as a
turning point in placing the epistemic agent's point of view centre stage. After all,
was not Kant the philosopher who with his self-styled 'Copernican revolution' re-
aligned philosophy around the human agents (as Copernicus re-aligned planetary
motion around the Sun)? Was not Kant the philosopher who clearly warned against
the sceptical threat facing anyone who asks how our representation of things conform to these things as they are in themselves? (see Kant 1781/1787, Bxx).

But while Kant certainly placed the human agent centre stage, he did not give
precise instructions as to how to 'exit' one's own perspective. How is it possible to
identify one's mode of knowledge as a particular perspective if one cannot exit it
and encounter others who occupy different perspectives? How could one recognise
one's own standpoint as such without a plurality of other possible standpoints? This
is the central question that Rachel Zuckert addresses in Chap. 1. Zuckert argues that
there is an inevitable tension inherent in the very idea of knowledge from a human
point of view. Kant maintains that one can only gain knowledge of the world from
within the human perspective. Yet the recognition of this fact requires one to be able
to step outside the human perspective and to acknowledge the existence of other
perspectives, which Kant seems to deny. Zuckert defends Kant's view from potential incoherency charges by examining Kant's Transcendental Dialectic in the
Critique of Pure Reason. Zuckert argues that reason with its ideas delivers a conception of a thing that cannot be presented in experience. Attempts to exit the human
perspective, and failures to do so (as Kant explored them in the Transcendental
Dialectic), can lead one to recognise the specificity and the limitations of the human
perspective, without ever being able to step outside it.

But maybe more than Kant himself, the Western philosopher who has more
clearly advocated a view known as perspectivism is Friedrich Nietzsche. In Chap. 2,
Steven D. Hales explores Nietzsche's two-tier perspectivism as encompassing a
first-order epistemic theory that takes truth as perspectival and a second-order methodological perspectivism aimed at enhancing 'understanding'. Hales defends
Nietzsche' 'positive epistemology' by responding to both critics who perceive
Nietzsche as a sceptic and those who have interpreted him as a pragmatist. He sur-
veys contemporary epistemological accounts concerned with the notion of 'under-

---

[1] The qualification of Western philosophy is important here because our already very selective
introduction to the topic will be confined to the three main figures of Kant, Nietzsche, and American
Pragmatism (with no implication or suggestion that similar themes cannot be found in other
Western authors, of course). A more comprehensive analysis would also need to include Arabic,
Indian, and Chinese philosophy (among others) where the notion of knowledge from a human
point of view might be cast in an interestingly new light. This would be an extensive scholarly
project to undertake on some another occasion, we hope.

standing' to motivate Nietzsche's methodological perspectivism and points out that the adoption of different perspectives, including erroneous ones, can further 'understanding'.

The perspectivalist line of inquiry that begins with Kant and continues with Nietzsche finds its mature expression in the multifaceted reflections of the movement known as American Pragmatism. Matthew Brown in Chap. 3 highlights how the notion of knowledge from a human point of view acquires wider resonance in the work of the American Pragmatists. Starting with the Pragmatist notions of inquiry and truth and potential lessons for perspectivists, Brown carves a path through a voluminous literature and analyses pluralistic metaphysics in the pragmatist tradition. His inquiry reveals certain shortcomings for perspectivism, such as a potential collapse into relativism, or a narrow Eurocentric focus in science. Brown suggests that these shortcomings can be overcome if perspectivists are willing to integrate certain lessons on truth, reality, and plurality from the American pragmatists. For example, the perspectivist can avail herself of Pierce's dynamic idea of *community of inquiry* to forgo the static and passive *vision metaphor*. Or learn from Addams' and Du Bois's standpoint theory to integrate a wider range of perspectives in science.

Unsurprisingly, American Pragmatism played a key role in informing one of the most influential contemporary advocates of a view closely related to perspectivism: Hilary Putnam's internal realism (or, as he later rebranded it 'realism with a human face'). In Chap. 4, Mario De Caro discusses Putnam's philosophical thoughts on reality and knowledge, and in particular his evolving views on what form of realism might be tenable. De Caro starts his survey with Putnam's views on physicalism and his criticism of metaphysical realism. He then turns to Putnam's internal realism, which, according to De Caro, was in part inspired by Kant, Peirce, and Dummett, and motivated by a renewed effort to respond to metaphysical realism. Putnam eventually abandoned internal realism in favour of 'liberal naturalism', a view that De Caro sees as congenial to Massimi's own version of perspectival realism.

These first four chapters set the historical stage for the second part of the book where the discussion switches to the ramifications of perspectivism in contemporary epistemology. What is it at stake in the seemingly anodyne claim that knowledge is 'from a human point of view'? In Chap. 5, Natalie Ashton looks at the topic through the lenses of contemporary feminist standpoint theory. She argues that both perspectivism and feminist standpoint theory have a lot to learn from relativism, as well as from one another. Ashton identifies elements of relativism at play in Ron Giere's perspectivism and in standpoint theory, respectively, and argues that there is an innocuous version of relativism that can benefit both views. One mistake that both Giere's perspectivism and feminist standpoint theories make, in Ashton's view, is to interpret relativism as asserting equal validity. The latter maintains that all rankings of different perspectives are equally correct, when in fact both views share with relativism the idea of non-neutrality, i.e. system-independent rankings are not possible. Ashton believes that once perspectivism and feminist standpoint theory embrace some version of non-silly relativism, both views will be better equipped to occupy the feasible middle-ground they are striving for.

In Chap. 6, Kareem Khalifa and Jared Millson put forward a view which they call 'inquisitive-truth monism', according to which it is not only true beliefs that are of epistemic value, but true answers to relevant questions. According to Khalifa and Millson, it is an inquirer's perspective that determines what questions are relevant, where the inquirer's perspective encompasses their interests, social role, and background assumptions. Khalifa and Millson's main motivation in pursuing inquisitive-truth monism is—in their own words—to account for 'the complexity of epistemically valuable undertakings characterizing the scientific endeavor'. They argue that traditional accounts, which focus on the acquisition of true beliefs, are inadequate to capture such complexity. They nod to perspectivism as a way of cashing out an alternative notion of epistemic normativity centred on the epistemic agent's perspectival interests. Along similar lines, Nick Treanor, in Chap. 7, undertakes an examination of epistemic normativity that takes perspectivism seriously. Treanor starts with a discussion of a widespread view about epistemic normativity that takes truth as a key norm for beliefs. On this view, shared by Alvin Goldman and Ernest Sosa among others, to know is to believe the truth—as much truth as is possible—and avoid error. Treanor highlights problems with this conception of epistemic normativity, focused as it is on more true and less false beliefs, and suggests a different way of thinking about epistemic normativity and a perspectival challenge looming in the horizon.

Sosa's epistemological view is also the starting point for Adam Carter's analysis in Chap. 8. Carter focuses on Sosa's 'virtue perspectivism' as a two-tier epistemological stance, whereby the reliability of first-order animal knowledge requires an ascent to second-order reflective or perspectival knowledge. Despite its success at averting scepticism and regress, critics have however lamented that virtue perspectivism falls prey of circular strategies. Carter's aim in this chapter is to tease out the criticisms and defend Sosa's virtue perspectivism from circularity-based objections levelled at the view by Barry Stroud, Baron Reed, and Richard Fumerton.

Aptly, this edited collection concludes with Chap. 9 by Barry Stroud himself, who undertakes a conceptual analysis of the very notion of 'knowledge from a human point of view'. By investigating the ways in which human beings come to know and what it means for one to come to know something, Stroud addresses the sceptical challenges to the possibility of knowledge and general concerns about knowledge and truth that a perspectival realist might have. He argues that to occupy a "human point of view" is to be fully engaged in the community of human knowers and to be committed to the world's being the way it is widely known to be. However, he also warns that this way of thinking about the original question does not have anything distinctively perspectival. And that maybe a better way of understanding the notion of 'knowledge from a human point of view' is to reflect not directly on human knowledge as such, but on human beings, 'their regarding themselves as enquirers or knowers'. Like Treanor, Stroud too invites 'aspiring perspectivists' to ask themselves questions about

> what we primarily want to understand about the acquisition and development of what we call human knowledge. Is it human acceptance—and rejection—of more and more theories or hypotheses that we think needs accounting for? Or is it the fact of theory change, or the

competition among theories: how can we tell which is best? Or is what we want to account for the progressive accumulation of more and more of what we call human knowledge. (Stroud, Chap 9)

These pressing questions remain ongoing concerns for aspiring perspectivists. Barry Stroud sadly and untimely passed away since writing this Chapter. We dedicate this volume to his memory, and hope this edited collection will prompt more and broader reflections on a fast-growing topic with a long-standing philosophical history.

Edinburgh, UK   Ana-Maria Crețu
October 2019   Michela Massimi

Contents

Contents

About the Editors and Contributors

Editors

Ana-Maria Crețu is a postdoctoral researcher working within the ERC project *Perspectival Realism. Science, Knowledge, and Truth from a Human Vantage Point* at the University of Edinburgh. Her research is principally within history and philosophy of science, with a particular emphasis on scientific classifications, real patterns, and disagreements in science.

Michela Massimi is Professor of Philosophy of Science at the University of Edinburgh. Her research interests are in the philosophy of science, the history and philosophy of modern physics, and Kant's philosophy of nature. She is the author of *Pauli's Exclusion Principle* (CUP, 2005) and co-editor of *Kant and the Laws of Nature* (Cambridge University Press, 2017) and *Understanding Perspectivism* (Routledge, 2019). She is the PI on the ERC project *Perspectival Realism.*

Contributors

Natalie Alana Ashton is a postdoctoral researcher working within the ERC project *The Emergence of Relativism* at the University of Vienna. She works on issues relating to justification, scepticism, and relativism in both traditional and feminist epistemology, and she has published a number of papers on these issues.

Matthew J. Brown is Associate Professor of Philosophy and History of Ideas at the University of Texas at Dallas and the Director of the Center for Values in Medicine, Science, and Technology. He has published extensively on a range of topics from science and society to cognitive science, and the history of philosophy.

Adam Carter is Lecturer in Philosophy at the University of Glasgow. He is the author of *Metaepistemology and Relativism* (Palgrave Macmillan, 2016) and is working on a forthcoming book with Clayton Littlejohn entitled *This Is Epistemology* (Wiley-Blackwell).

Mario De Caro is Professor of Moral Philosophy at Rome Tre, and a regular Visiting Professor at Tufts University. He is the editor, with David Macarthur, of the volumes *Naturalism in Question* (Harvard University Press, 2004) and *Naturalism and Normativity* (Columbia University Press, 2010) and of two volumes of essays by Hilary Putnam: *Philosophy in the Age of Science* (with D. Macarthur, Harvard University Press, 2012) and *Naturalism, Realism, and Normativity* (Columbia University Press, 2016).

Steven D. Hales is Professor and Chair in the Department of Philosophy at Bloomsburg University. He specialises in epistemology and metaphysics and has co-edited books on both Nietzsche's philosophy and on relativism. He is the author of *Nietzsche's Perspectivism*, with Rex Welshon (University of Illinois Press, 2000), and of *Relativism and the Foundations of Philosophy* (MIT, 2006).

Kareem Khalifa is Professor at Middlebury College. His research focuses on issues in general philosophy of science, philosophy of social science, and epistemology, areas in which he has published extensively. He is the author of *Understanding, Explanation, and Scientific Knowledge* (CUP, 2017).

Jared Millson is a Visiting Assistant Professor of Philosophy and Co-Chair in the Department of Philosophy at Agnes Scott College. He has published a number of journal articles and book chapters on logic, philosophy of language, and philosophy of science, and he is currently writing a book on theories of scientific explanation with Kareem Khalifa and Mark Risjord.

Barry Stroud was Willis S. and Marion Slusser Professor Emeritus of Philosophy in the Philosophy Department at the University of California, Berkeley. He was the winner of the Matchette Prize (1979) for his book *Hume* (Routledge, 1977) and the author of *The Significance of Philosophical Scepticism* (OUP, 1984), *The Quest for Reality* (OUP, 2002), *Engagement and Metaphysical Dissatisfaction* (OUP, 2011), as well as four volumes of collected essays also published by Oxford University Press.

Nick Treanor is Reader in Philosophy at the University of Edinburgh. His research is primarily within metaphysics, epistemology, philosophy of mind, and philosophy of language, and he has published a number of papers and book chapters within these areas.

Rachel Zuckert is Professor of Philosophy at Northwestern University. She is an expert on Kant and German idealism and the author of *Kant on Beauty and Biology: An Interpretation of the Critique of Judgment* (CUP, 2017) and of *Herder's Naturalist Aesthetics* (CUP, 2019).

Chapter 1
Attempting to Exit the Human Perspective: A Priori Experimentation in Kant's *Critique of Pure Reason*

Rachel Zuckert

Abstract I consider a problem for Kant's transcendental idealism if one construes it as a claim that human beings know from a particular, specifically human perspective. Namely, ordinarily when we speak of someone seeing from a perspective, we understand other people to have other perspectives, and think that people can change their perspectives by moving away from them, to a different one. So one may recognize that one's own perspective is a perspective by comparing to others, by seeing a former perspective from a new vantage point. But Kant denies such plurality and variability for the perspective he identifies; it is the human perspective as such. Thus, one may worry that Kant's view. is incoherent: Kant claims that we can know only from one perspective, yet, in order to recognize that perspective, he himself must stand "outside" of it. I consider a potential Kantian response to this charge, in the form of an interpretation of the Dialectic section of the first *Critique*. When one attempts to know things that lie beyond the human perspective — to exit it — one falls into contradictions and empty thinking. These failed attempts to exit the human perspective constitute its horizon, a limit recognizable without one needing truly (but impossibly) to occupy a different perspective. Such failed attempts, I argue, are some of the confirming results of the a priori experimentation Kant proposes in the Preface to the *Critique*: his hypothesis of transcendental idealism is shown to identify the dividing line between successful and failed, productive and contradictory attempts at human knowledge.

Keywords A priori knowledge · Transcendental idealism · Kant's theoretical philosophy · Kant's critique of rationalist metaphysics · Philosophical methodology · Perspective

R. Zuckert (✉)
Northwestern University, Evanston, IL, USA
e-mail: r-zuckert@northwestern.edu

© The Author(s) 2020
A. Crețu, M. Massimi (eds.), *Knowledge from a Human Point of View*,
Synthese Library 416, https://doi.org/10.1007/978-3-030-27041-4_1

1.1 Introduction

As thematized in this volume, philosophers often speak of perspectives, meant not (usually) in the literal senses of painting technique or of viewer's perception of a segment of a visual field from a particular spatial position, but in the also familiar, though more metaphorical sense of having a "point of view", thinking of things in one's own way, framed by one's particular modes of attentiveness, organizing principles or interests, and so forth. In his *Critique of Pure Reason*, Kant uses the closely related term "standpoint" to describe the human epistemic condition. Summing up his argument that space is an a priori form of intuition, a framework for human sensibility, and so formative of objects as they appear to such sensibility, he writes:

> We can accordingly speak of space, extended beings, and so on only from the human *standpoint [Standpunkt]*. If we depart from the subjective condition under which alone we can acquire outer intuition, namely that through which we may be affected by objects, then the representation of space signifies nothing at all.[1]

Here Kant comes quite close to saying that human knowledge – at least knowledge of sensibly presented objects – is inevitably from, formed by, a human perspective. More generally, one could use this metaphor to gloss Kant's central philosophical doctrine, transcendental idealism: that human "cognition reaches appearances [i.e., objects as they appear to us, in sensible experience] only, leaving the thing in itself as something actual for itself but uncognized by us" (Bxx). One might say, then, that on Kant's view human beings can know, even with necessity, how things will be from our perspective – i.e., the spatio-temporal realm of appearances – but *only* within and for our own perspective.

My paper concerns a question brought out by this way of characterizing Kant's position, specifically by a disanalogy between it and our ordinary way of thinking of perspectives. Ordinarily we take perspectives to be plural and variable: different people have different perspectives; individuals can also change their standpoints and thereby look (from "outside") at their own previous perspective, recognizing it as such.[2] But, at least with respect to the human standpoint centrally at issue in his transcendental idealism, Kant denies this.[3] There are and can be no plural perspectives among human beings, nor can any individual move away from the standpoint she occupies. For the human standpoint as such is both universal, shared by all human beings, and necessary for each individual human knower.[4] Thus one may

---

[1] Kant, *Critique of Pure Reason*, A26/B42, my emphasis. Citations to the *Critique* will be, as customary, to the A/B page numbers of the first and second editions. Translations are from Kant (1998).

[2] On perspective as a metaphor used in philosophy, I have profited from Conant (2005), as well as Moore (1997), though the latter is a deeper treatment than I can properly engage with here.

[3] Kant does not deny, of course, that individual human beings could have different empirical standpoints (either literal or metaphorical).

[4] This disanalogy is heightened by the fact that Kant uses "standpoint" to describe *space*, so that the very idea of moving to a different spatial position, or of different people occupying different spatial positions, is ruled out. (Here Kant follows Leibniz, who takes space to be merely phenomenal, yet describes the truly-existing monads as having their own distinctive "perspectives" on the world-

ask: what does it mean to identify one's mode of knowledge as a particular perspective if one cannot exit it, if one does not encounter others who occupy different ones? How could one recognize one's own standpoint as such without such plurality or possibility to exit?

These questions do not, I think, result merely from pressing a metaphorical expression. Rather, they reformulate a central question concerning the Kantian philosophical enterprise, raised in various forms from early on in Kant's reception. Hegel and other post-Kantians accuse Kant of self-contradiction for similar reasons: Kant claims that human beings can know only from one perspective, yet it would seem that in order to recognize that perspective as such, he does and must stand outside of it.

In this paper, I investigate one Kantian response to such concerns: in brief, the Dialectic section of the *Critique*. In the Dialectic, Kant portrays rationalist metaphysics as a failed attempt to know things beyond experience, and I shall suggest that these failed metaphysical views can be understood as attempts to exit the human perspective; one discovers, in the course of the Dialectic, that these attempts lead to cognitive failure. Thus, on Kant's view, we do not have to inhabit another perspective or – impossibly – stand outside our own perspective to establish its limits. The limits of our perspective can be established from within, from the epistemic problems that arise when we attempt to transcend it. I begin with a slightly more extensive discussion of the problem, before turning to propose this view of the Dialectic.

1.2 The Problem, in Some More Detail

As just noted, when we speak of someone occupying a particular standpoint or having a particular perspective in everyday life, we take such perspectives to be plural and variable: we understand other people to have other perspectives, from their own, different spatial locations, and that one can change one's perspective, moving away to another (location or, by metaphorical extension, attitude or theoretical position). One may recognize that one has a specific perspective then, either by recognizing that others see things differently or by varying one's own position, looking back at one's original perspective, and so seeing its location and limitation. It seems prima facie unclear, then, what a "perspective" (in this everyday sense) would mean in cases of universal, unaltered, perhaps unalterable agreement: *are* there "perspectives" on well-established mathematical theorems or basic facts such as 'human beings need to eat to survive'? How would one establish that this agreement is actually a universally shared perspective? By comparison to what?

---

whole.) In line with Strawson's objections to Kant (1966, e.g., p. 41), and with Conant (2005), one might accuse Kant therefore of illicitly (perhaps even self-contradictorily) extending an intramural experiential concept (of occupying a spatial position) to extra-experiential use. I will not be able to take up this specific version of the problem here.

These questions are, of course, the ones I wish to press about Kant's claim to identify a universally shared, necessary human perspective on objects of experience, Kant's famous philosophical "Copernican Revolution". According to this transcendental idealist "altered method of our way of thinking", Kant writes, he can explain the possibility of a priori knowledge: "we can cognize of things a priori only what we ourselves have put into them" (Bxviii). More precisely, we can know objects a priori insofar as they "conform to the constitution of our faculty of intuition", and therefore "the *experience* in which alone they can be cognized ... conforms to [certain a priori] concepts" (Bxvii). Kant elaborates this approach in the Transcendental Aesthetic and Analytic sections of the *Critique*: we *can* know a priori about objects of experience, concerning both their spatio-temporal character – their conformity to the a priori conditions of human sensibility – and their conformity to the categories, the a priori concepts of the understanding (such as unity, negation, cause and substance). Thus, Kant claims, we can know that "nature (in the empirical sense) [is] ... the combination of appearances ... in accordance with law ... indeed in accord with its original laws, in accordance with which experience itself first becomes possible" (A216/B263). Correspondingly, Kant argues, these claims – what we can know a priori – are correct from the human standpoint alone. They concern objects as appearing *to us*, given the nature of human cognition, and otherwise – as he strongly puts it concerning space (at A26/B42, quoted above) – are "nothing".

Henceforward, I will refer to this complex of doctrines as "the human perspective" on Kant's view. Namely, (1) human beings must employ together the cognitive faculties of understanding and sensibility – must use both concepts and sensible intuitions – to know objects; (2) each of these human cognitive faculties furnishes and necessarily relies upon a priori cognitive representations (the categories and space/time, respectively); (3) objects known by such human beings – the objects that (as it were) come into view for such a perspective – are correspondingly and therefore law-governed, spatio-temporal objects of experience, or "appearances". These three aspects of the human perspective may be glossed in turn as describing (1) the subject who knows, (3) the object known thereby, and (2) the constitutive structure of (1) that is carried over to or in some way determinative of (3). Perhaps one could therefore call (2) the structure of the perspective (i.e., it describes how the subject is oriented to objects, and correspondingly how objects will appear to or come into view for that subject). And again, the question I aim to raise concerning this complex of doctrines is: given the universal character of this "standpoint", the necessity for every human knower that she uses these cognitive capacities to know objects, in what sense is this a perspective, and how can it be identified as one?

I note first that Kant is well aware of the philosophical-methodological utility of invoking plurality or variability of perspectives, to become aware of one's own perspective as such. In the Paralogisms chapter of the first *Critique*, Kant explicitly invokes the conception of an observer at a different "standpoint" in order to make clear that the constant (potential) presence of self-consciousness in one's experience, the persisting identity of one's representation of one's "I", does not prove that one's self is a persistent substantial entity.[5] Kant here proposes an alternative perspective – or stages a "thought-experiment" of inhabiting another perspective – in order to make one aware that certain facts (the persistence of one's "I") *are* features of one's own perspective, and do not necessarily hold of things independently of that perspective. Earlier in the Paralogisms, Kant also brings out the difficulty with this methodology if deployed to try to identify necessary, shared features of a human perspective as such: "It is obvious", he writes, "that if one wants to represent a thinking being, one must put oneself in its place, and thus substitute one's own subject for the object one wants to consider (which is not the case in any other species of investigation)".[6] If there are necessary features of the human mode of knowing as such, then any attempt by a human subject to inhabit a different perspective from that one, to look at it from outside, will bring those necessary features along with it. One will "substitute" oneself, place oneself, along with all the necessary features of human knowing, within that alternative perspective.

Thus, I suggest, the ordinary way of making sense of, and recognizing, a "perspective" or "standpoint" appears to be ruled out for the purported Kantian human standpoint – and Kant acknowledges as much. But he does, I think, propose (at least) two ways of identifying the human perspective in the Transcendental Aesthetic and Analytic.

First, Kant extends the notion of plurality of perspectives to refer to other, non-human beings. He suggests that the necessary, universally shared human perspective may be distinguished from, and recognized by comparison to, that of possible other finite intellects that might have other forms of intuition (other than space and time),[7] or, more prominently, a divine mode of knowing: intuitive intellect or intellectual intuition.[8] These distinctions seem to me pedagogically helpful (as it were) for bringing readers to understand the kind of position Kant is proposing, to see that he wishes to argue that human beings have particular modes of access to

---

[5] A 362-3. Kant uses *"Gesichtspunkt"* here, and then glosses the same observer position as *"Standpunkt"* at A364.

[6] A 353-4. "In its place" translates *"an seine Stelle"*. These passages may seem to conflict with one another, since in the first, Kant suggests that one *can* regard a thinker from outside, precisely not "substituting one's own subject" for that thinker's self-consciousness. But we may note that in the thought experiment the outside observer is furnished with the same necessary features of human knowledge (specifically, time as a form of sensibility but also, presumably, its own self-consciousness). Thus, the observer does not in that respect occupy an "external" (i.e., non-human) perspective. The staged outside observer also does not aim to judge "my self" as a knower, but in "outer intuition", that is, as a substantial thing persisting in time, which supports or underlies that thinking or (in short) as object.

[7] See A27/B43, following the characterization of space as belonging to the human standpoint.

[8] There has been discussion about whether intuitive intellect and intellectual intuition are the same – e.g., Förster (2009) – but I think the present line of argument is independent of a decision on this issue.

the world, or (in Henry Allison's terms) "epistemic conditions" governing our approach to objects.[9]

The gesture to other potential modes of knowing makes the general shape of this proposal clear. But these comparisons do seem to me to be mere gestures. By themselves, they are not decisive for establishing that space, time, and the categories – the lawfulness of nature – are factual merely for, and derived from, the human perspective. For, in Kantian terms, it is unclear whether these purported alternative perspectives are "real possibilities": though we can (somewhat) imagine such entities and their possible alternative modes of knowing, we do not know that there really could be such alternative modes of knowing. Or, in the everyday terms I sketched above: our inability really to conceive of such intellects, of what their modes of knowing would be like, of how things could be from another, entirely different sort of perspective, would follow again from the inescapability of the universal, human perspective. If we think in a serious way about thinking (knowing) beings, we will "substitute our own subjects" for such beings. Why, in short, should we take it that there really could be such other "perspectives"?

Of course one can always consider whether there might not be some radically other way of looking at the world, whether one's beliefs, modes of thinking and so forth might be mistaken. So too can one wonder whether there might be radically other sorts of beings, with radically other modes of knowing. But unless something further is said, these considerations are somewhat idle – perhaps a salutary reminder to be epistemically humble, but not providing significant ground for restricting one's claims, demoting them (as it were) to a mere perspective.

Second, Kant provides more substantive, direct arguments that space and time are and must be conceived as forms of human intuition in the Transcendental Aesthetic. Likewise (if more complicatedly), he argues in the Transcendental Deduction that objects of experience are constituted as such by the categories – that, in other words, such objects are substances with attributes, subject to causally governed changes, and so forth – on the grounds that otherwise *human* experience would be impossible. Thus, Kant concludes, objects of experience may legitimately be so characterized only from a human perspective, only under the restriction that we are discussing objects as presented to human knowers.

These lines of argument are too complex for me to discuss in any detail. I will therefore just sketch a worry – loosely inspired by (a perhaps odd combination of) Strawson and Heidegger[10] – about taking them as decisively identifying a *human perspective* as such, i.e., as entailing that the cognitive claims established are local to, or even just descriptive (solely) of some *particular* sort of knower and objects as known thereby. That is, these Kantian lines of argument might be consistent with

---

[9] I think it is not coincidental that Allison's interpretation, which emphasizes the epistemological (not metaphysical) dimensions of transcendental idealism – and so is close to thinking of Kant as identifying a human "perspective" – also emphasizes the Kantian contrast between human and (purported) divine cognition (see Allison 1983, chapter 2, especially 19–24).

[10] Heidegger (1990) and Strawson (1966). Obviously, I am presenting a very broad-brush picture here.

the conclusion that this sort of object – natural, sensibly intuitable, temporal objects – must be so, for *any* knower who would be acquainted with them.[11] Therefore, we need not say that these objects only appear so, or that they are so only for a specific sort of knower. Rather, any knower able to know about such objects will need to be furnished with the appropriate cognitive capacities – here a priori forms of intuition (space and time), and a priori concepts such as substance or cause – in order to recognize those facts about those objects.[12] Kant's restriction of such claims to a human standpoint might then seem like an over-interpretation, an unjustified specification of the results he has in fact achieved (from analysis of the character of objects of experience). Certain concepts and principles may apply to a restricted range of objects, or to objects only under certain descriptions – objects of sensible nature, say – but it may not be clear exactly why one needs to add that such objects are such only from or particularly for the human perspective.[13]

There are many possible responses to this sort of worry. Kant himself seems inclined to (what we now call) an argument to the best explanation. That is, if one grants that human beings are furnished with a priori representations (that is, the second item on the list of three points defining the human perspective above, what I called the "structure" of the perspective) and that objects must conform to such a priori forms in order to "appear" to us – so Kant argues – it would be a miraculous coincidence that those objects *also* just happen to be so, independently of our cognitive demands, our modes of apprehending them, their modes of appearing to us. The more plausible, efficient explanation is that they are so in virtue of appearing to us (see Kant 1772). The recent renewed interest in metaphysical readings of transcendental idealism, particularly of Kant's doctrines concerning sensibility, could also be seen as responses of the following sort: Kant's arguments concerning space and time establish that all spatio-temporal objects simply cannot be things in themselves, given the (dependent, relational, or otherwise metaphysically questionable)

---

[11] Unsurprisingly, Heidegger and Strawson get to this conclusion differently, Strawson proposing a Kantian "metaphysics of experience" (analysis of what objects must be like, if there is to be experience), while Heidegger takes Kant to engage in a metaphysical ("ontological") investigation of human subjectivity, which also reveals the ontology of this sort of object. Heidegger does emphasize the finitude of the subject, a Kantian theme connected to the topic of this paper, but finitude need not be understood in terms of the perspectival limitation of claims, with which I am concerned. For a recent realist interpretation of Heidegger and his reception of Kant (somewhat along these lines), see Kinkaid (2018).

[12] I take it that this could describe a contemporary naturalist approach to a priori knowledge: human beings justifiably deploy a priori concepts (innate cognitive modes of interpreting experience such as object-permanence) because this is accurate to the world – and this "fit" between innate capacities and world is in turn to be explained in evolutionary terms (that human beings have evolved cognitively to fit their environment). Of course such a line of argument is subject to Nietzsche's objection that falsities might be evolutionarily just as useful to the human animal as accurate conceptualization.

[13] The "neglected alternative" objection (prosecuted by Adolf Trendelenburg in several works in the 1860s), namely that Kant does not rule out the possibility that space and time are *both* forms of our intuition *and* characterize things in themselves, occupies similar territory. I do not think the approach I explore here can address this way of formulating the problem.

nature of space and time. Space and time must be understood to exist only in their relations to us; knowledge claims concerning them (or objects "in" them) hold only to or from a human standpoint (see, notably, Langton 1998 and Allais 2015). Here I explore a different possible response, namely the Dialectic of the *Critique of Pure Reason*.

1.3   The Transcendental Dialectic: An Alternative Perspective

The Transcendental Dialectic comprises, roughly, the second half of the *Critique of Pure Reason*.[14] Here Kant is concerned with what he takes to be the central subject matters of traditional "special" metaphysics: soul, God, and world-whole. He aims to explain how human beings come to formulate conceptions of these "transcendent" entities – i.e., things that transcend human sensible experience – and to show definitively that philosophers' claims to know a priori about such entities are mistaken. In particular, he devotes the Paralogisms chapter (quoted above) to a priori arguments that purport to elucidate the nature of the soul, as simple, immaterial substance. Kant objects that all such arguments are either tautological – based upon and merely re-describing formal features of (the representation of) self-consciousness, but no entity that purportedly underlies it – or invalid (insofar as they do claim to establish truths about some such entity, they illicitly import contents from sensible experience). The Antinomies chapter concerns a priori responses to questions concerning the world-whole: most famously, is it thoroughly deterministic, or is there a place for freedom? But also, is there a beginning of time, an end of space? Is there a smallest, most basic component of material objects, or are they infinitely divisible? Is there a necessary being grounding all the contingent elements of the world, or is it turtles all the way down? Kant argues that there are, always and systematically, two conflicting answers to such questions, both of which are supported by a priori rational argument. In pursuing such questions, therefore, we fall into a "contradiction of reason" (A408/B435). Finally, Kant devotes the last chapter of the Dialectic, entitled "The Ideal", to explaining the origin of the idea of God and to arguing that a priori proofs of God's existence fail, most famously because existence is "not a real predicate" (A598/B626).

---

[14] According to Kant's own divisions of the work, this is not accurate: organizationally (if not in terms of page numbers), the Transcendental Dialectic is something like an eighth of the work. For Kant divides the book into two (unequal) halves: the Method, a section at the end of the work concerning what philosophy is to do after Kantian critique, and the Elements, the much longer first half of the work, which includes the Transcendental Aesthetic, containing Kant's arguments that space and time are a priori forms of intuition, and the Transcendental Logic. The Logic is in turn divided into two subsections, the Transcendental Analytic, concerning the a priori concepts of the understanding (categories) appropriately and necessarily employed with respect to objects of experience, and the Transcendental Dialectic, concerning the ideas of reason that purport to present objects beyond experience.

These arguments contribute, in obvious and direct ways, to Kant's philosophical enterprise in the *Critique*. Kant announces that this work concerns the possibility of metaphysics (Axii, Bxv); it therefore appropriately includes consideration of rationalist special metaphysics. This discussion belongs as well to Kant's project of "critique", i.e., of reason's "self-knowledge" (Axi-xii). For Kant argues that reason formulates the ideas of soul, God, and world-whole as a result of its "drive" to attain complete, satisfactory, systematic explanation, or (in his terms) to find the "unconditioned", the ultimate totality of conditions from which all (experiential or other) truths could be derived, on which all experiential things could be grounded.[15] And of course insofar as Kant shows that we cannot know about soul, God, or world-whole by pure reason, independently of experience, he supports his transcendental idealist conclusion concerning human knowledge as such, i.e., that it is restricted to objects of experience.[16]

But in the B Preface, Kant claims that the results of the Dialectic not only support his *restriction* of human knowledge to objects of experience, but also serve as a "splendid touchstone of" his characterization of the a priori knowledge that we *do* have (on his view) *as* characteristic of the human perspective – as comprising "what we ourselves have put into" objects of experience (Bxviii). For, Kant proposes, one might see the *Critique* as an "experiment".[17] The hypothesis tested is his own transcendental idealism, namely (again) that human a priori knowledge can be explained if we take objects of experience to "conform to the constitution of our faculty of intuition" so that "the *experience* in which alone they can be cognized … conforms to [certain a priori] concepts" (Bxvii). By contrast, Kant writes, the arguments treated in the Dialectic concern "objects" not considered as appearances, but "insofar as they are thought merely through reason".[18] In showing that those arguments fall into "contradiction" (or otherwise fail), Kant claims, he also confirms his hypothesis: "The experiment decides for the correctness of [the] distinction" between appearances and things in themselves, thus for his construal of objects of experience *as* appearances, i.e., as objects for or within the human perspective alone (Bxix).

---

[15] Bxx; "drive" translates "*treibt*". See also A327/B383-4 and A332/B389.

[16] One might wonder whether the field of potential non-experiential knowledge is larger than rationalist specialist metaphysics: in showing that we cannot know *these things* (soul, world-whole, God) a priori, has Kant thereby shown that we cannot know anything about *anything* non-experiential? Here Kant's transcendental-psychological etiology of the ideas of reason proves crucial: it aims to show that these are the only rational (non-practical) ideas human beings are able to form of transcendent things. Investigation of this line of thought lies beyond the scope of this paper, however.

[17] Bxvi. One suggestive element of this description of Kant's project is that it proposes a new view of Kant's philosophical methodology: as experimental, trying out hypotheses, seeing which best fits the philosophical "facts", rather than as the "apodictic" demonstrative argument to which Kant lays claim (but which have not been found so decisive by most of his readers).

[18] Bxviii. Kant's contrast in this immediate textual location is to the knowledge we have of objects in accord with "what we ourselves have put into them", whereas I use a more specific contrast (to "intuition" and "experience") from earlier in the passage. I return to Kant's broader contrast below.

It is this proposed, indirect function of the Dialectic – to confirm the conclusions in the Analytic – that I shall explore here.[19] For one may well wonder, why and how does *not* being able to know about immortal souls, God, or the world-whole show that empirical-causal explanations or geometrical proofs and the like *are* knowledge from or for the human perspective alone? I will propose that the Dialectic – the "experimental" results Kant attains there – may be seen as an answer, of a kind, to the problem I sketched above. That is, I shall suggest that in the Dialectic, Kant portrays human knowers as, in fact, exiting the human perspective, *sort of*: rationalist metaphysicians *attempt*, but ultimately fail, to exit the human perspective.[20] Thus the Dialectic portrays an alternative perspective from which one can regard one's original perspective and recognize it as such – though it also ultimately affirms that the human perspective is the only successful one for human knowers. To be clear, I propose that Kant's view is not that any other perspective is incoherent, ultimately inconceivable; the alternative perspective is conceivable, even "inhabitable" by the human subject (at least the philosophically minded one). But our attempts at knowledge from within it turn out to fail. I will now discuss the two sides of this proposal in turn – the way in which the portrayed metaphysicians exit the human perspective, and thereby recognize it as such, and then the qualifications on that alternative perspective (why I characterize it above as a mere "attempt", or a "sort of" alternative).

1.4   An Alternative Perspective

As just quoted, Kant describes rationalist special metaphysics as treating objects not as appearances, but as "thought merely through reason" (Bxviii). I propose that this brief description may be understood as referring to an alternative human perspective, taken up by the rationalist metaphysician, and portrayed or even enacted in Kant's presentation of arguments concerning soul, world or God, in the voice (as it

---

[19] I should note that the subject matters of the Dialectic also identify the "meaning" of Kant's assertion in the Analytic that human cognition is only from the human perspective, in another sense: it *matters* that objects of experience are such only for the human perspective, so as, famously, to "make room for faith" (Bxxx). If objects of sensible experience are (mere) appearances, then one may believe – perhaps has to believe – that sensible nature is not all there is, that empirical scientific description does not exhaust truths about things. In particular, Kant claims, we are permitted to believe that our agency may be free and that God exists – as we are required to do for moral reasons. Hence the desirability of the *limitation* connoted by identifying a perspective, even one shared by all human beings.

Given that morality requires this different view of oneself, one might think that *it* provides an alternative perspective from which we can recognize the limits of the human (cognitive) perspective (I owe this suggestion to Alix Cohen). This, however, seems to me to mistake the structure of Kant's philosophical position: as clearly stated in the "make room for faith" passage, theoretical (cognitive) limits must first be established, on theoretical grounds, in order to permit moral self-reconception as an alternative perspective.

[20] Here I qualify the claims of the preceding section that insofar as human beings share the same standpoint, we are incapable of exiting it or occupying an alternative one.

were) of the metaphysician. In this alternative perspective (corresponding to the three points concerning "the human perspective", above), (1) the human knower employs the cognitive faculties of understanding and reason, not sensibility; (2) therein uses the categories (a priori concepts of the understanding) to attempt to describe objects of the a priori ideas of reason, i.e., (3) objects (soul, world-whole, and God) that are non-sensible and so transcendent, definitively beyond or outside of sensible experience. In particular, because these ideas are formulated by reason, their objects are understood as totalities or as the "unconditioned". That is, each idea of reason (of soul, world-whole, or God) identifies a purported ultimate and total ground for some aspect of experience – whether the subject or an aspect (spatio-temporal or conceptual) of known objects (see A333-4/B390-91). In so doing, reason formulates a conception of a thing that cannot be presented in experience, which never presents ultimate grounds, nor complete totalities (at least that we could know or recognize as such). I will henceforth refer to *this* complex as "the alternative perspective".

This alternative perspective can, I suggest, allow the human knower (and Kant describing the same) to identify "the human perspective" *as* a perspective, as a specific – not unique – way to approach objects. One may recognize that using concepts to judge sensibly given objects is not the only form knowledge may take if one also can do something else, i.e., think of objects using pure reason. In particular, the rationalist metaphysician who attempts to know about the soul by pure reason takes herself to have intellectual intuition (and so not to require sensible intuition to present objects about which to think). For she takes it that her self-consciousness, her non-sensible (intellectual) representation of "I think", presents her *immediately* with a particular, *given* entity – herself, her soul.[21] In Kant's treatment of the project of proving God's existence, the rationalist metaphysician again attempts to perform a cognitive function Kant denies to the human perspective: to "synthesize" concepts with one another directly (and justifiably), without recourse to sensible experience as a ground for linking the two concepts. For, Kant maintains, in arguing a priori that God exists, rational theologians aim to connect two, non-identical concepts, both attained by considering what the "unconditioned condition" for contingent beings might be: the concept of necessarily existing being and the concept of the being with the "highest reality" (*ens realissimum*).[22]

---

[21] This is how I read Kant's references to intuition in the Paralogisms, e.g., at B411-12, though this is a more controversial interpretive and reconstructive claim than I can defend here. I note also that my characterization of this representation as a purported intellectual intuition – an immediate presentation of a particular (object), given by intellect, not sensibility – does not include an element often associated (by Kant and by scholars) with intellectual intuition, namely that such an intellect would produce its objects. I think these two aspects of purported intellectual intuition can in fact be separated: both characterize God's purported intellect, on Kant's sketchy characterization, but it is not clear to me why non-divine intellects could not have a different form of intuitive intellection. So Förster (2009) also concludes, on different grounds.

[22] Insofar as Kant's portrayed rationalist metaphysician seems to perform (somewhat) different cognitive actions in the three chapters of the Dialectic, one might think that there are three (or more) alternative perspectives portrayed in the Dialectic. I do not have a worked-out account of

In thus subtracting sensibility and laying claim instead to different cognitive abilities (to synthesize concepts and intellectual intuition), these two rationalist metaphysical projects illuminate that the human perspective *is* a specific perspective, deploying specific cognitive faculties/types of representations: understanding together with sensibility, concepts as applied to and synthesizing sensibly given intuitions. Since these are, moreover, cognitive attitudes that human beings may take, projects in which human knowers may engage, in which philosophers in the history of philosophy *have* engaged – and which Kant enacts in the Dialectic – they make good on the gesture (in the Analytic) at possible alternatives to the human perspective.

Kant's discussion in the Antinomies chapter is more obviously concerned with the failures of the alternative perspective (to which I turn shortly). But the position taken up by the rationalist metaphysicians here also, arguably, illuminates why one might take the human perspective not just to be a specific perspective, but also a limited one, which gives a view of "mere" appearances, not self-standing, independent things. The alternative perspective portrayed in the Paralogisms and Ideal chapters may already appear superior to the human perspective, in that the cognition therein enacted appears independent of sensibility, and thus both more "native" to human thinking (more centrally part of who we are as knowers, less dependent on external information) and immune to experiential refutation. But in the Antinomies chapter, this superiority is more directly at issue. For here the rationalist metaphysician (the alternative perspective) focuses on the very objects of experience, treating them as subject to reason's demands to specify their ultimate conditions (their spatio-temporal extent, their ultimate substantial components, their causal and existential foundations). This demand brings out the insufficiency of objects of experience by the standards of reason: none of them will ever count as such an ultimate ground, all are ineliminably contingent, dependent on something else. In order to take them to have fully grounded existence, all must either be conceived of as parts of an infinite (and so not-experienceable) whole or as furnished with a further, ultimate, non-experiential grounding, in a different sort of object (God as first cause and necessary being, or perhaps monads—see A416/B443-A418/B446). Consequently, the human perspective, in focusing exclusively upon objects of experience, is concerned exclusively with objects of lesser, dependent metaphysical status – things that by their very nature depend upon something outside that perspective for their "real", ultimate, grounded being, that are not (in this sense) things in themselves.[23]

---

how all three are deeply connected, but I think my general point is unchanged either way – some "experimental" alternative fails.

[23] To be clear, the metaphysical status – of dependency or contingency – at which I gesture here is not dependence on human knowers (i.e., it is not idealism). Rather, these objects are dependent (on other things, of some other, less flimsy kind). Insofar as these are the only objects known by the human perspective, then, that perspective does not provide a view of metaphysically independent things. As I discuss in the Coda, it would require further argument to move from this dependence to idealism (dependence of appearances on human knowers/perspective).

1.5   Only Sort of an Alternative Perspective

One may immediately object the following, however: can it be that on Kant's view, human beings actually *can* occupy the alternative perspective? Can it be, as just suggested, that this alternative perspective is somehow *superior* to the human perspective, in furnishing knowledge of what things in themselves are like? Kant's locutions occasionally suggest so. In the B Preface passage concerning the experiment, for example, he contrasts the human perspective and its *limited* knowledge – knowledge (only) of "what we ourselves have put into" objects – with the conception of objects by pure reason. The latter (Kant implies) *does not* comprise knowledge of what we have "put into" objects, rather it conceives them in themselves.[24]

Yet this cannot quite be right, given Kant's emphatic denial of our knowledge of things in themselves, his restriction of human knowledge to the human perspective. And of course in the Dialectic Kant not only enacts the rationalist metaphysical project – the alternative perspective – but also, indeed dominantly, argues that it is a failure. This is the principal reason for my qualifications (above): the alternative perspective is only *sort of* an alternative perspective. It is not fully a perspective on objects, first, because its cognitive attempts are failures, its point of view empty, fragmentary, contradictory. And, I add now, Kant's arguments against its attempts at knowledge articulate the ways in which it fails such that the human knower is, as it were, pointed back to the human perspective.

That is, in his critical analysis both of the proofs for the existence of God and arguments concerning the nature of the soul, Kant argues in effect that the alternative perspective fails in its attempts to know its supersensible objects precisely because it excludes sensibility – precisely in the way in which it attempts to exit the human perspective. On Kant's analysis, the attempted proofs of God's existence all succumb to conceptual fragmentation: for human knowers, there is no way, justifiably, to synthesize two different concepts, unless one can invoke sensible experience. But of course the alternative perspective – especially when attempting to know God as necessary being – excludes sensibility.[25] Hence, Kant claims, there remains an ineliminable gulf between the concepts to be synthesized.[26] As it turns out, the alternative perspective does not present a coherent view of a unified object.

---

[24] Bxviii. One might call this the "residual rationalism" in Kant's thought: *if* we were to know things in themselves, we *would* know them by reason alone. Perhaps he thinks that insofar as we have a positive conception of things as separate from experience (not just the "negative" noumenon of B307, i.e., an abstraction from objects of experience), this conception is gained from reason (in the form of the ideas of reason). But I think he also believes that reason (not sensibility) articulates the independent-ness of things in themselves. Langton (1998) is still the most worked-out account (to my knowledge) of Kantian idealism as a form of such residual rationalism.

[25] I gloss over considerable detail here, and proceed blithely on the assumption that Kant's criticisms of rationalist metaphysics are successful.

[26] So I construe Kant's arguments (e.g., A611/B639-A613/B640) that we cannot prove that the necessarily existing being is the *ens realissimum* or vice versa.

The rationalist metaphysical claim to know the self by intellectual intuition is likewise false, Kant argues: insofar as there is some intuitive – immediate, existential, singular – content to the rationalists' representation of their selves, this content comes from temporally formed sensibility, not from awareness of the "I" of thought (itself merely an empty, formal structure of self-consciousness).[27] Kant thus accuses the rationalist metaphysical arguments of the fallacy of ambiguity: their purported conclusions, their purported grasp of a supersensible object, turns on unacknowledged substitution of sensible for purportedly purely intellectual conceptual content. In the terms I have been using, the alternative perspective fails to know its object (engaging rather in tautology or fallacy), and it proves to be dependent (in an unacknowledged way) on the human perspective, specifically on sensibility; it is dependent, then, precisely on that which it claims to exclude.

The failure of the alternative perspective is even more centrally at issue in the Antinomies chapter, where metaphysical reason is portrayed as enmeshed in unavoidable contradictions. Contradiction, of course, is a cognitive failure; if one is led to assert both p and not-p, something has gone wrong, by reason's own most basic standard. A perspective plagued by contradiction is clearly a failed perspective; it cannot provide an integrated "view" of or approach to objects.

Kant provides a more complicated explanation of this failure than in the two other cases. The antinomies do not arise because of the exclusion of sensibility, for here the alternative perspective attempts to find its own proper objects, things conceived by reason, in or at the basis of sensibly presented objects of experience. But there are two ways to perform such an identification of sensible and rationally conceived objects. As in the antithesis positions – which Kant also names "dogmatic empiricism" (A470-71/B498-9) – one may insist that spatio-temporal objects of experience *are* fully self-standing things, as conceived by reason.[28] Or, with more traditional rationalists (the thesis positions), one may claim that spatio-temporal objects must be grounded in some more basic, rationally conceived things. Hence, necessarily, two opposed positions – the antinomial conflict – as well as Kant's resolution to it: if one distinguishes objects of experience from objects thought by reason, the contradiction disappears.[29] Indeed Kant claims more strongly that one learns here to distinguish appearances from things in themselves, i.e., to accept the

---

[27] See A350-1: because the representation of the "I" is merely a form of thought, not an intuition, it does not have content such as "everlasting duration" (which would be content of sensible, i.e., temporal intuition and which is introduced in thinking of the self as enduring substance).

[28] This position may not seem to conceive of objects by pure reason, but insofar as it takes objects of experience to be infinite (or parts of an infinite series), it re-describes them in terms that are themselves neither drawn from, nor presentable within sensible experience. For such reasons, Boehm (2014) argues that Kant could be targeting Spinoza in the antithesis positions.

[29] Kant proffers two different sorts of resolutions (to the first and second, and to the third and fourth antinomies, respectively): either one stops demanding that appearances conform to reason's standards of what independent things would be (because they are merely appearances) or one sees that one might be able to assert both p and not-p, but this would no longer be a contradiction, because the two claims concern different (aspects of) things.

restriction of human knowledge to the human perspective as such (Bxx; I return briefly to this characterization of the results of the experiment in the Coda).

The experiment that Kant enacts in the Dialectic thus shows what happens when one tries to exit the human perspective: cognitive failure. We must therefore qualify – but not dismiss – my above characterization of the alternative perspective: it is not, fully or stably, an alternative perspective accessible to or inhabitable by human beings. However, I propose that it remains a "position" from which one can recognize the specific, limited character of the actual human perspective, universally shared and necessary as it may be.[30] In trying and failing to know oneself through intellectual intuition, or to know God by synthesizing concepts, one recognizes that one *is specifically* and ineluctably dependent on sensibility, as an element of cognition distinct from thought. And in recognizing (through antinomial contradictions) that the sensible objects with which the human perspective is concerned do not sustain the demands of full rational explanation – that they are not, and cannot be derived from unconditioned, absolute grounds – one recognizes the restrictedness of the human perspective, its insufficiency to answer all questions about, to provide full grounding for, the only objects it does know. Thus, though there is no *actual* variability or plurality of (successful) human cognitive standpoints on Kant's view, nonetheless human beings can attempt to exit the human perspective. And through engaging in such attempts, and recognizing their failures – through philosophical experimentation – one can recognize the specificity and the limitations of the human perspective.[31]

1.6   Coda

In closing, I'd like to acknowledge two limitations of the above proposals. I have suggested that Kant's philosophical experiment can address the question how one can identify one's own, necessary perspective, without really being able to leave it, not (incoherently or inconsistently) from outside but from within. There is another important aspect of this project that I have not discussed explicitly, however, which

---

[30] Here one might differentiate my proposal from the following stronger, narrower construal of the experiment and its results: the experiment concerns the antinomies alone, particularly the first antinomy, and it establishes that the alternative perspective therein taken up is contradictory, completely incoherent. This interpretation could be supported textually, and might be appealing in presenting Kant as arguing more directly for his view that space and time are merely parts of the human perspective (for it would take the experiment to show that thinking otherwise is flatly incoherent). But it also, I think problematically, isolates the antinomies from the project of the Dialectic (and Appendix to the Dialectic) as a whole. The experiment would also not be able to perform the (on my view methodologically important) role of comparison case: a completely incoherent position cannot be inhabited, so could not be one from which another perspective could be recognized as such.

[31] As Mark Alznauer noted (in personal communication), one may then ask: is this recognition true from or for the human perspective (alone)?

may seem all the same to take one outside the human perspective. That is, Kant's explanations of the origins of the ideas of reason (soul, God, and world-whole) and his diagnoses of where the alternative perspective goes wrong in trying to know them – at all of which I have merely gestured here – are also crucial to the "experiment" I am describing. In these explanations, Kant aims to show that the alternative perspective is, as it were, constructed out of materials (such as the structure of self-consciousness or the serial relations of conditioned to condition that connect objects in experience) constitutive of the human perspective, or by purporting to subtract some such materials (i.e., primarily, sensibility). It is in this way as well not really an *alternative* perspective – not really other, but rather a newly (and, Kant wishes us to discover, badly) reorganized version of the human perspective.[32] But there is one signal element in these explanations and in the aspirations of the alternative perspective, which is not such a material: reason itself. To identify the human perspective as such, to see its limits, dependence, and insufficiency, human beings must have a faculty that, as Kant writes, "has a natural propensity to overstep all these boundaries" (A642/B670), that demands absolute explanations, grasp of totalities and necessary grounds, that drives toward the unconditioned, and so (according to Kant) pursues illusory, deceptive self-transcendence. But is this self-transcending drive really accommodated within the human perspective?

Perhaps an answer to this question could be found in Kant's account in the Appendix to the Transcendental Dialectic of reason as providing regulative principles for empirical investigation, and indeed of the ideas of reason as opening up a "focus imaginarius" – aspirational end points or goals that orient or unify the human perspective in a way similar to the focal points of literal, visual perspectives (on this account, see Massimi 2017, 2018; Zuckert 2017). In other words, it appears that for Kant, if appropriately directed or understood, reason and its self-transcendence are constitutive of the limited, human perspective as such. Perhaps so – but then again one may wonder what it can mean to have a standpoint, a perspective, defined by an orientation to something beyond itself.

Second, I hope to have suggested not just how Kant can explain that one is aware of the human perspective (despite its universality and necessity), but also the

---

[32] Though I do not have space to elaborate, I think Kant does and must aim to show this: he must explain how rationalist metaphysics is even possible as a project, given his view of the actual character and limits of human cognition. I am inclined, moreover, to think that this part of Kant's project in the Dialectic could explain the sometimes problematic nature of Kant's arguments. That is, Kant occasionally seems not just to use his own terminology, but to invoke some of his own substantive commitments (as established in the Analytic) in his treatment of the rationalist metaphysical project. Such use of his own doctrines seems to beg the question with respect to what one might call his primary aim in the Dialectic: to argue that (contra the rationalists) human cognition is limited in the Kantian way, Kant cannot use as premises those very claims concerning human cognitive limitations. (This problem seems to me to plague the interpretation in Grier 2001 as well, though this fine book has been crucial for my understanding of the Dialectic.) But for this *secondary* aim – explaining how limited human knowers could attempt to transcend their own, necessary perspective – Kant's own understanding of the human perspective must be central. Kant seems to me not always clearly to disambiguate these two argumentative tasks; many of his criticisms of rationalist metaphysics might consequently need to be reformulated so as not to beg the question.

meaningfulness of taking the human perspective *as* a perspective: it is a specific form of cognition, of a specific kind of object, that may be taken as, according to the standards of reason, insufficient, and limited. None of this amounts, however, to showing that the objects of such cognition are metaphysically or otherwise dependent on the human perspective, nor that claims about them are true only for, localized to, that perspective. As noted above, however, Kant claims that his experiment does show this:

> on the assumption that our cognition from experience conforms to the objects as things in themselves [i.e., as carried out by the alternative perspective in the Antinomies], the unconditioned **cannot be thought at all without contradiction,** but … if we assume that our representation of things as they are given to us does not conform to these things as they are in themselves but rather that these objects as appearances conform to our way of representing, then **the contradiction disappears.** (Bxx)

On Kant's description, the contradictions in the antinomies do not show (or do not merely show) that an alternative *perspective*, that of rationalist special metaphysics, fails. Rather, Kant claims that the conflicts arise if one denies the distinction between appearances and things in themselves, that is, denies *that* there is a human perspective at all. Thus, he famously claims, the antinomies are an "indirect proof" of transcendental idealism: the contradiction disappears if one recognizes that appearances are appearances, i.e., *only* so for the human perspective. This claim raises questions for my description of the alternative perspective *as* a perspective. For example, does this re-description (as I hope) free Kant from his residual-rationalist, apparently dogmatic claim that reason (what I call the alternative perspective) would provide human beings with knowledge of things in themselves? Or is the alternative perspective insufficiently robust to delimit the human perspective as such, unless it is taken so to transcend perspectival knowledge altogether?

Acknowledgments For comments on previous drafts of the paper, I am grateful to Mark Alznauer, Karl Ameriks, Morganna Lambeth, Katalin Makkai and participants in the "Knowledge from a Human Point of View" conference at the University of Edinburgh, especially Lorenzo Spagnesi and Michela Massimi.

Bibliography

Allais, L. (2015). *Manifest reality: Kant's idealism and his realism.* Oxford: Oxford University Press.

Allison, H. (1983). *Kant's transcendental idealism: An interpretation and defense.* New Haven: Yale University Press.

Boehm, O. (2014). *Kant's critique of Spinoza.* Oxford: Oxford University Press.

Conant, J. (2005). The dialectic of perspectivism, I. *Sats: Nordic Journal of Philosophy, 6*(2), 5–50.

Förster, E. (2009). The significance of §76 and 77 of the *Critique of Judgment* for post-Kantian philosophy, part I. *Graduate Faculty Philosophy Journal, 30*(2), 1–21.

Grier, M. (2001). *Kant's doctrine of transcendental illusion.* Cambridge: Cambridge University Press.

Heidegger, M. (1990). *Kant and the problem of metaphysics* (Richard Taft, Trans.). Bloomington: Indiana University Press.

Kant, I. (1772). Letter to Markus Herz of February 12. Reproduced in *Gesammelte Schriften* 20, 129–35. Berlin: De Gruyter.

Kant, I. (1998). *Critique of pure reason* (Paul Guyer and Allen Wood, Trans.). Cambridge: Cambridge University Press.

Kinkaid, J. (2018). Phenomenology, idealism, and the legacy of Kant. *British Journal of the History of Philosophy, 13*(2), 1–22.

Langton, R. (1998). *Kantian humility and things in themselves*. Oxford: Oxford University Press.

Massimi, M. (2017). What is this thing called 'scientific knowledge'? Kant on imaginary standpoints and the regulative role of reason. *Kant Yearbook, 9*, 63–83.

Massimi, M. (2018). Points of View: Kant on Perspectival Knowledge. *Synthese*, special issue on the current relevance of Kant's philosophy. Online first.

Moore, A. W. (1997). *Points of view*. Oxford: Clarendon Press.

Strawson, P. F. (1966). *The bounds of sense: An essay on Kant's Critique of Pure Reason*. New York and London: Routledge.

Zuckert, R. (2017). Kantian ideas of reason and empirical scientific investigation. In M. Massimi & A. Breitenbach (Eds.), *Kant and the laws of nature* (pp. 89–107). Cambridge: Cambridge University Press.

Chapter 2
Nietzsche's Epistemic Perspectivism

Steven D. Hales

Abstract Nietzsche offers a positive epistemology, and those who interpret him as a skeptic or a mere pragmatist are mistaken. Instead he supports what he calls *perspectivism*. This is a familiar take on Nietzsche, as perspectivism has been analyzed by many previous interpreters. The present paper presents a sketch of the textually best supported and logically most consistent treatment of perspectivism as a first-order epistemic theory. What's original in the present paper is an argument that Nietzsche also offers a second-order methodological perspectivism aimed at enhancing *understanding*, an epistemic state distinct from knowledge. Just as Descartes considers and rejects radical skepticism while at the same time adopting methodological skepticism, one could consistently reject perspectivism as a theory of knowledge while accepting it as contributing to our understanding. It is argued that Nietzsche's perspectivism is in fact two-tiered: knowledge is perspectival because truth itself is, and in addition there is a methodological perspectivism in which distinct ways of knowing are utilized to produce understanding. A review of the manner in which understanding is conceptualized in contemporary epistemology and philosophy of science serves to illuminate how Nietzsche was tackling these ideas.

Keywords Nietzsche · Perspectivism · Understanding · Knowledge

2.1 Introduction

In this paper I will argue that Nietzsche offers a positive epistemology, and that those who interpret him as a sceptic or a mere pragmatist are mistaken. Instead he supports what he calls *perspectivism*. So far this is not a new take on Nietzsche, as perspectivism has been analyzed by many previous interpreters. I present a sketch of what I think is textually best supported and logically most consistent treatment of perspectivism as a first-order epistemic theory. What is new in the present paper is

S. D. Hales (✉)
Department of Philosophy, Bloomsburg University, Bloomsburg, PA, USA
e-mail: hales@bloomu.edu

A. Crețu, M. Massimi (eds.), *Knowledge from a Human Point of View*,
Synthese Library 416, https://doi.org/10.1007/978-3-030-27041-4_2

an argument that Nietzsche also offers a second-order methodological perspectivism aimed at enhancing *understanding*, an epistemic state distinct from knowledge. One could consistently reject perspectivism as a theory of knowledge while accepting it as contributing to our understanding. In fact, I believe he accepts both.

One thing Nietzsche loves is the hyperbolic smackdown followed by (partly) walking it back. He uses this tactic on practically every topic he touches. In *Beyond Good and Evil* (*BGE* 108) and *Twilight of the Idols* (*TI* VII:1) Nietzsche declares that there are no moral facts whatsoever and then goes on to enunciate formulas for greatness and recipes for virtue (*Ecce Homo* II:10; *AC* 11).[1] He denounces philosophers' obsession with the will (*BGE* 19) and then promotes the will to power.

Nietzsche's audacious decrees are usually meant to shock the reader out of a complacent conformity to traditional dogmas. By artificially turning up the contrast, he casts rival ideas into high relief and highlights the alternative ideas that he is proposing. Nietzsche is fond of dismantling philosophical structures, taking their components and re-using them in new ways. He says as much in *Assorted Opinions and Maxims* (*AOM* 201): "The philosopher believes that the value of his philosophy lies in the whole, in the building: posterity discovers it in the bricks with which he built and which are then often used again for better building: in the fact, that is to say, that that building can be destroyed and *nonetheless* possess value as material." Of course, Nietzsche's own positive views are famously hard to pin down, since he approaches "deep problems like cold baths: quickly into them and quickly out again" (*The Gay Science* 381). The matter of his epistemic perspectivism is no exception, as he touches on it, moves on, and then returns to it many pages or even books later.

---

[1] *HATH = Human, All Too Human*, ed. and trans. Marion Faber and Stephen Lehmann (Lincoln: University of Nebraska Press, 1984; original edition: 1878).

AOM = Assorted Opinions and Maxims, trans. R.J. Hollingdale (Cambridge: Cambridge University Press, 1986; original edition: 1879)

D = Daybreak, trans. R.J. Hollingdale (Cambridge: Cambridge University Press, 1982; original edition: 1881).

GS = The Gay Science, ed. and trans. Walter Kaufmann (New York: Vintage Books, 1974; original edition: 1882).

Z = Thus Spoke Zarathustra, ed. and trans. Walter Kaufmann, in *The Portable Nietzsche* (New York: Viking Penguin, 1954; original edition: 1885).

BGE = Beyond Good and Evil, ed. and trans. Walter Kaufmann (New York: Vintage Books, 1966; original edition: 1886).

GM = On the Genealogy of Morals, ed. and trans. Walter Kaufmann (New York: Vintage Books, 1967; original edition: 1887).

AC = The Antichrist, ed. and trans. R.J. Hollingdale (New York: Viking Penguin, 1968; original edition: 1895).

TI = The Twilight of the Idols, ed. and trans. R.J. Hollingdale (New York: Viking Penguin, 1968; original edition: 1889).

EH = Ecce Homo, ed. and trans. Walter Kaufmann (New York: Vintage Books, 1967; original edition: 1908).

WP = The Will to Power, ed. and trans. Walter Kaufmann and R.J. Hollingdale (New York: Vintage Books, 1968).

Nietzsche's blitzkrieg is not as successful a rhetorical strategy as he might have hoped. It is far too easy to latch onto some portion of his work and treat it as a synecdoche that represents the whole. The result is a secondary literature that contains only pockets of agreement and little satisfactory unification. Nietzsche worried a good bit about being misunderstood (*GM* III 1, *AOM* 137, *EH* preface 1), while also impishly admitting that "I obviously do everything to be 'hard to understand' myself!" (*BGE* 27). It's hard to tell what sort of readers Nietzsche was hoping for, since he certainly had no optimism that philosophers were going to figure it all out. As he puts it in *The Gay Science* (§333), "*Conscious* thinking, especially that of the philosopher, is the least vigorous and therefore also the relatively mildest and calmest form of thinking; and thus precisely philosophers are most apt to be led astray about the nature of knowledge."[2]

2.2 The Skeptical and Pragmatic Interpretations

In practice, the result is that when Nietzsche writes things like "the biggest fable of all is the fable of knowledge" (*The Will to Power* 555); "delusion and error are conditions of human knowledge and sensation" (*GS* 107); "there are *no eternal facts*, nor are there any absolute truths" (*Human, All Too Human* 2), or when he repeatedly hammers against the sensibility of Kantian things-in-themselves that could be the objects of knowledge, as he does in the fourth chapter of *Twilight of the Idols*, many readers assume that he is an epistemological skeptic.

Babette Babich, for example, writes, "Nietzsche contends that knowledge — even as a limited perspective — is impossible" (Babich 1994, 80). R.J. Hollingdale concurs. He claims that according to Nietzsche, "[i]n the sense in which philosophers are accustomed to use the word ... there is *no* knowledge" (Hollingdale 1973, 131). Jean Granier also agrees that for Nietzsche "the traditional concept of knowledge appears as a pseudo-concept" (Granier 1977, 192). Alan Schrift thinks that Nietzsche rejects epistemology entirely as a result of his skepticism: "[he] does not provide a theory at all; it is a rhetorical strategy that offers an alternative to the traditional epistemological conception of knowledge as the possession of some stable, eternal 'entities', whether these be considered 'truths', 'facts', 'meanings', 'propositions', or whatever .... Nietzsche views these 'entities' as beyond the limits of human comprehension, and ... he concludes that we are surely incapable of 'knowing' them" (Schrift 1990, 145). Peter Poellner sees Nietzsche as part of a Cartesian skeptical tradition, where "even if we were able to rationally justify a thought ... we would have no good reason to regard it as true" (Poellner 1995, 63). Willard Mittelman sums up the skeptical interpretation: "for Nietzsche knowledge of the world is impossible" (Mittelman 1984, 8).

---

[2] As Socrates asked Theaetetus, "do you fancy it is a small matter to discover the nature of knowledge? Is it not one of the hardest questions?" *Theaetetus* 148.c.

Jessica Berry offers a subtle analysis of Nietzsche as belonging to the Pyrrhonian skeptical tradition, something quite distinct from the more common readings of him as a Cartesian skeptic. For ancient skeptics like Pyrrho of Elis and Sextus Empiricus, the peace of mind they termed *ataraxia* was to be achieved through suspending judgment about everything that is less than self-evident. Arguments can be mounted for and against any propositions, making them uncertain and inconclusive, leaving the inquirer indefinitely indecisive and unsure what to believe. For the Pyrrhonian skeptic, that is the proper frame of mind. Berry holds that Nietzsche's perspectivism is primarily epistemic — he recognizes a variety of perspectives as equipollent, which leads to the Pyrrhonian suspension of judgment and a rejection of dogmatic views that take one perspective as superior to another. Perspectivism as a doctrine or positive position of any kind would be just another sort of dogmatism on her reading (Berry 2011, chapter 4).

It is puzzling that those who regard Nietzsche as a skeptic not only skip over his criticisms of skepticism, but somehow miss the occasions when he walks his more extreme claims about knowledge back. In *BGE* 208 Nietzsche remarks that "when a philosopher suggests these days that he is not a skeptic ... everybody is annoyed". Why are they annoyed? Is it because skepticism is so clearly right that only retrograde, dogmatic troglodytes still claim to have knowledge? No, skeptics are "delicate" and "timid"; they suffer from "nervous exhaustion" and "sickliness." Skepticism, Nietzsche claims, is a fashionable decadence, a soporific sedative; "entertaining no hypotheses at all might well be part of good taste". In other words, the rejection of skepticism is a social *faux pas* in fashionably effete circles, whereas the claim of knowledge — *that* is dangerous, an explosive rumbling in the distance. In the same section Nietzsche equates objective knowledge with dressed-up skepticism, which he regards as mere epistemic trendiness. But he is not dismissing knowledge; in the same breath that he denounces skepticism, Nietzsche rejects absolutism as a kindred crime. Skepticism and absolutism are opposite sides of the same devalued coin.

Nietzsche does not sentence knowledge — in some sense of the term — to the gallows. Consider that "no honey is sweeter than that of knowledge" (*HATH* 292) and his assertion that "whoever seriously wants to become free ... his will desires nothing more urgently than knowledge and the means to it — that is, the enduring condition in which he is best able to engage in knowledge" (*HATH* 288). He also connects knowledge with both pleasure and happiness: "Why is knowledge, the element of researchers and philosophers, linked to pleasure? First and foremost, because by it we gain awareness of our power ... Second, because, as we gain knowledge, we surpass older ideas and their representatives, become victors, or at least believe ourselves to be. Third, because any new knowledge, however small, makes us feel superior to *everyone* and unique in understanding this matter correctly" (*HATH* 252). Likewise, happiness is positively undesirable without knowledge: "our drive for knowledge has become too strong for us to be able to want happiness without knowledge ... Knowledge has in us been transformed into a passion which shrinks at no sacrifice and at bottom fears nothing but its own extinction" (*Daybreak* 429). Nor is the happiness provided by knowledge cheap or ephemeral: rather, "the happiness of the man of knowledge enhances the beauty of the world and makes all that exists sunnier; knowledge casts its beauty not only over things but in the long run into things ..." (*D* 550).

The quotations in the preceding paragraph are from *Human, All Too Human* and *Daybreak*, two of Nietzsche's earlier works, and those enamored of the idea that Nietzsche's thought progressed through distinct phases (like Maudemarie Clark and Brian Leiter) might be inclined to regard those passages as examples of an early, "positivistic" phase, rather than of his mature thinking. It is true that Nietzsche did change his mind about some topics. Richard Wagner is the most obvious, where the hagiography of *The Birth of Tragedy* gave way to the vitriol of *Nietzsche Contra Wagner* and *The Case of Wagner*. However, a close examination of Nietzsche's writings from throughout his life demonstrates that he displays continuing respect for knowledge. Even in *The Gay Science*, the source of much of Nietzsche's epistemological critique, he characterizes himself as a lover of knowledge (*GS* 14), a seeker of knowledge (*GS* 380), and as someone greedy for knowledge (*GS* 242, 249). The most famous announcement of *The Gay Science*, the death of God, is taken to be a welcome epistemic harbinger; now that God is dead, "all the daring of the lover of knowledge is permitted again; the sea, *our* sea, lies open again; perhaps there has never been such an 'open sea'" (*GS* 343). Nietzsche is clear in *The Gay Science* (§324) that this is cause for rejoicing: "and knowledge itself ... for me it is a world of dangers and victories in which heroic feelings, too, find places to dance and play. '*Life as a means to knowledge*' — with this principle in one's heart one can live not only boldly but even gaily, and laugh gaily too."

In *Thus Spoke Zarathustra* one finds Zarathustra saying "With knowledge the body purifies itself; making experiments with knowledge it elevates itself; in the lover of knowledge all instincts become holy; in the elevated, the soul becomes gay" (*Z* I:22.2). In the first section of the preface to *On the Genealogy of Morals* Nietzsche writes that the treasure for "we men of knowledge" is "where the beehives of knowledge are". This sort of praise for knowledge continued to the end of his productive life. In *The Antichrist* (48, see also 49) he criticizes Christianity on the grounds that it opposes science and knowledge, characterizing knowledge as "emancipation from the priest"; and in *Ecce Homo* ("The Birth of Tragedy": 2) he writes, "Knowledge, saying Yes to reality, is just as necessary for the strong as cowardice and the flight from reality — as the 'ideal' is for the weak, who are inspired by weakness".

So Nietzsche has some kind of positive attitude towards and program regarding knowledge and a program. But what is it? *Beyond Good and Evil* opens with him questioning the value of truth, and why its pursuit should be seen as the highest value. "Knowledge for its own sake" is a form of unexamined moralizing (*BGE* 63). Nietzsche often writes about the practicality of beliefs as a distinct virtue from their truth. Consider *The Gay Science* 354: "we simply lack any organ for knowledge, for 'truth': we 'know' (or believe or imagine) just as much as may be *useful* in the interests of the human herd, the species ...". In the same passage, Nietzsche goes so far as to speculate about conditions in which we would be so severely mistaken about what is useful for us that it would cause human extinction. In *The Will to Power* (493) he writes that "truth is the kind of error without which a certain species of life could not live", and suggests in *WP* (515) that our goal is "not 'to know' but to schematize — to impose upon chaos as much regularity and form as our practical needs require". In *BGE* 4 he writes, "the falseness of a judgment is for us not necessarily an objection ... the question is to what extent it is life-promoting, life-preserving, species-preserving, perhaps even species-cultivating". We need to recognize untruth as a condition of life (*BGE* 4).

It is not surprising that these sorts of passages led some interpreters to conclude that Nietzsche was a pragmatist, where knowledge just consists in accepting those doctrines that are helpful or productive for our lives. Arthur Danto led the way, writing "Nietzsche ... advanced a pragmatic criterion of truth" (Danto 1965, 72), and he was followed by Reudiger Grimm, who writes, "the criterion to be met by any of these perspectival 'errors' [that Nietzsche discusses] is not one of veracity, but rather one of utility. In point of usefulness there is a great deal of difference between interpretations ..." (Grimm 1977, 70). More recently, Tsarina Doyle interprets Nietzsche as defending *internal realism*, a view that Hilary Putnam regarded as the heir to the American pragmatist tradition (Doyle 2009, chapter 2). Neil Sinhababu is even more explicit: "Nietzschean pragmatism is the view that one should believe whatever best promotes life, even things that are untrue ..." (Sinhababu 2017, 56).

2.3   Nietzsche's First-Order Perspectivism

Nietzsche's pragmatist interpreters have a case to be made, but like the skeptics' reading, it is incomplete, as we will see. Probably the most popular analysis of his epistemology is that he defends *perspectivism*, a sort of relativism about truth and knowledge that has tendrils throughout his work. Perspectivism in contemporary philosophy of science has to do with human limitations and focused interest. For example, we see colors only along a very narrow band of electromagnetic radiation, which restricts our connection to the external world. In addition, our scientific theories often aim to model the world only at some interest-relative scale, as we see moving from atomic physics to chemistry to biology. Scientific perspectivism of this sort is arguably a realist theory. Nietzsche's perspectivism is broader and less focused. For him, perspectivism is not one precisely defined doctrine, but a cluster of related ideas about the subjectivity of truth, anti-realist metaphysics, a bundle theory of objects, the revaluation of values and the creation of one's own virtues, and the role of varying interpretations in knowledge (Hales and Welshon 2000). It is the latter that I will focus on here.

Recall that in *HATH* 2 Nietzsche writes that "there are *no eternal facts*, nor are there any absolute truths". The gripping thing about this passage is that Nietzsche is plainly not denying the existence of *truth*, but the existence of *absolute* truths or *eternal* facts. He elaborates on this theme in *The Genealogy of Morals* (III:12), which is worth quoting at length:

Let us be on guard against the dangerous old conceptual fiction that posited a 'pure, will-less, painless, timeless knowing subject'; let us guard against the snares of such contradic-tory concepts as 'pure reason', 'absolute spirituality', 'knowledge in itself': these always demand that we should think of an eye that is completely unthinkable, an eye turned in no particular direction, in which the active and interpreting forces, through which alone seeing becomes seeing *something*, are supposed to be lacking; these always demand of the eye an absurdity and a nonsense. There is *only* a perspective seeing, *only* a perspective knowing; and the *more* affects we allow to speak about one thing, the *more* eyes, different eyes, we can use to observe one thing, the more complete will our 'concept' of this thing, our 'objec-tivity' be.

Plato and Kant are the obvious targets of Nietzsche's criticisms. Plato's forms were the ideal, self-instantiating paradigms of concepts, the perfect examples of right-ness, beauty, and goodness (*Parmenides* 130b). Our apprehension of those ideas through the muddled testimony of the senses is inadequate for true knowledge, and it is only by recollection of the forms prior to birth that we can properly grasp them. Kant's doctrine of things-in-themselves similarly posits unmediated objects that exist in a noumenal realm beyond the ken of human apprehension, and our knowl-edge of those things (such as it is) occurs only after filtering through the categories of experience. In both cases the "real world" is unattainable, the God's-eye view tantalizingly out of reach. Human knowledge is a pale simulacrum of true contact with reality.

For Nietzsche, Plato and Kant were united by their otherworldliness — both the metaphysical view that reality is something beyond or other than the empirical world of sensation and the epistemological view that we see that world through a glass, darkly. "Pure reason" Nietzsche dismisses as a falsification of the senses, a means of telling ourselves that the true nature of the world is fixed and unchanging. He accuses philosophers of conceptual idolatry, dehistoricizing our categories and concepts and treating them as unalterably eternal. Kant's pure concepts of the understanding and Plato's ideal forms he regards as examples of this "rude fetish-ism" (*TI*, "'Reason' in Philosophy" 5). Nietzsche regards modern Christianity, with its belief in an eternal afterlife beyond our earthly experience and an abstract notion of an immutable God, as of a piece with the metaphysics of Plato and Kant (cf. *AC* 16–19). It is this epistemology of the other — somehow more authentic or genu-ine — world offered by Christianity, Plato, and Kant that Nietzsche wishes to declare, at last, a myth (*TI* IV).

Instead of absolute truths and impartially objective knowledge of a supra-empirical world, Nietzsche offers a vision of partial, fragmentary, perspectival knowledge. His regular praise of Heraclitus's acceptance of flux and becoming instead of the Platonic longing for an unchanging invisible world suggests the epis-temology Nietzsche supports. We are not prevented from ideal cognition by our this-worldly empirical limitations; Nietzsche wants to abolish the *apparent world/ real world* dichotomy altogether. Rather, the very nature of truth, and therefore our knowledge of those truths, is in some way dependent on perspective. Nothing is true outside or independent of perspectives; the idea of extra-perspectival knowledge is too redolent of those Kantian or Platonic epistemologies that Nietzsche has already dismissed. As he comments in *Thus Spoke Zarathustra* (III:11.2), "'This is my way;

where is yours?'—thus I answer those who asked me 'the way'. For the way — that does not exist." There is also the testimony of *The Will to Power* (540), "There are many kinds of eyes. Even the sphinx has eyes — and consequently there are many kinds of 'truths', and consequently there is no truth." One doesn't simply know or not know a claim, because claims are not true or false without being embedded in a particular perspective.

Perspectives themselves are more than mere beliefs, an error that would make everyone infallible. If what's true for a person is a matter of his or her perspective, and a perspective is just his or her set of beliefs, then the believer could never be wrong. Fortunately, Nietzsche is aware of this, and makes explicit the distinction between being believed true in a perspective and being true in a perspective. In *The Antichrist* (23) he writes "truth and the belief that something is true: two completely diverse worlds of interest". Perspectives instead have something to do with centers of interest, or attitudes organized around a common concern: "all evaluation is made from a definite perspective: that of the preservation of the individual, a community, a race, a state, a church, a faith, a culture" (*WP* 259). Perspectives are both local and abstract, they are "the basic condition of all life" (*BGE* preface). Nietzsche is not precise about exactly what he considers to be a perspective, but they are best characterized as ways of knowing, or doxastic practices.

According to William Alston, a doxastic practice is "the exercise of a system or constellation of belief-forming habits or mechanisms, each realizing a function that yields beliefs with a certain kind of content form inputs of a certain type" (Alston 1991, 155). Doxastic practices for Alston are abstract types; concrete specific practices are tokens of those types. Alston is especially interested in what makes a particular doxastic practice Christian, and states that Christian practice "takes the Bible, the ecumenical councils of the undivided church, Christian experience through the ages, Christian thought, and more generally the Christian tradition as normative sources of its overrider system" (Alston 1991, 193). Similar reasoning would apply to a scientific perspective, or an aesthetic one. A scientific doxastic practice includes something like the norms of data collection and analysis, the methods of empirical observation and testing, the assumptions and puzzles of normal science, and an epistemological commitment to wide reflective equilibrium. It is only from within, or with reference to, such perspectives that anything could be said to be true.

Without a doubt, the most troublesome aspect of this reading of perspectivism is how the doctrine might be applied to itself. If *the* way to the truth does not exist, because there are many kinds of truths, then what shall we make of those very claims? Isn't perspectivism itself just another perspective? Isn't it supposed to be true for everyone that there are distinct interests that constitute the different perspectives? Writing in 2001, Bernard Reginster describes this paradox of perspectivism as dominating the previous 20 years of English-speaking Nietzsche scholarship (Reginster 2001), and certainly it has bothered Nietzsche's commentators going back at least to Danto. If perspectivism is a perspective, then there are perspectives in which statements are untrue only in a perspective; if perspectivism is not a perspective, then it is untrue that every statement is true in some perspectives and untrue in others (Danto 1965, 80).

One way to salvage Nietzsche's perspectivism is the following: instead of insisting that *everything is perspectival*, one could aver that *everything true is perspectivally true*. The vital difference between the two formulations can be brought out with an analogy. Compare *everything is possible* to *everything true is possibly true*. No one except the pathologically optimistic would defend the idea that *everything is possible*, but *everything true is possibly true* is so obvious as to hardly rate a comment. *Everything true is possibly true* allows the possibility that there are necessary truths that are true in all worlds and it permits that some truths are merely contingent ones that are true in some worlds but false in others. Analogously, *everything true is perspectivally true* is compatible with there being absolute truths that are true in all perspectives while also permitting that there are merely perspectival truths that are true in some perspectives and false in others. Nietzsche is then free to argue that there are perspectives, that truth is indexed to perspectives, that there is no such thing as truth outside of or independent from perspectives, and so on. Those structural claims are true in all perspectives, without risk of self-refutation.[3]

2.4   Nietzsche's Second-Order Perspectivism

All the discussion of perspectivism so far has been treating it as a first-order theory, where our knowledge is perspectival because truth itself is. This has been the (or at least a prominent) mainstream way to understand Nietzsche's perspectivism, and has been the way I have defended and presented it in the past. However, I think there is another way into his perspectivist thinking that has not been previously discussed in the literature and is an intriguing new lens though which we can see his work. That is to regard perspectivism not as specific theory of knowledge, or even a collection of theories, but as a *strategic methodology*. The right analogy here is to Descartes's discussion of skepticism.

In the first *Meditation*, Descartes is troubled by how often he has been wrong in the past. This is especially bothersome because some areas of human inquiry — geometry, for example — seem set on firm foundations from which further facts can be rigorously derived. Ordinary empirical knowledge seems much shakier. Descartes's worries lead to two further lines of reflection: the first is doubt about the possibility of any kind of empirical knowledge, as raised in his famous skeptical arguments. The first skeptical argument is the contention that there are no detectable differences between the wakeful perception of reality and a realistic dream. Since we cannot distinguish between dreaming and reality, we cannot be sure that the testimony of our senses provides us with a window into the truth. The second

---

[3] These ideas were first developed in (Hales and Welshon 2000). Lightbody 2010, footnote 1 writes of this view "I believe it is the only consistent perspectivist position possible." A more general discussion of approaches to self-refutation is in (Hales forthcoming).

skeptical argument imagines the possibility of an evil demon "of utmost power and cunning" who devotes all of his energies to our deception. Like witnessing the world's greatest illusionist, we can never trust our experiences as representing the way the world really is. Of course, Descartes argues later in the *Meditations* that we do have knowledge, and that these skeptical arguments ultimately fail.

Pretty much all later epistemologists have thought that it is Descartes's rebuttals that ultimately fail, but that doesn't matter here. The point is that Descartes raises skeptical concerns, takes them seriously, and finally rejects them. This sort of first-order skepticism was one of his two lines of investigation stemming from the recognition of human fallibility. The second, which Descartes does not reject, is *methodological* skepticism. His methodological skepticism is contained in his proposal that "I should hold back my assent from opinions which are not completely certain and indubitable just as carefully as I do from those which are patently false" (*Meditations* I 18). The method of doubt does not presume that there cannot be empirical knowledge or that the dream and demon arguments have any purchase at all. Rather, it sets out an approach to further inquiries, namely the psychologically challenging attempt to suspend judgment about everything that can be doubted, until proven otherwise.

Descartes rejects first-order skepticism as a positive epistemic theory while promoting skepticism as a second-order methodology. It is consistent to accept both or reject both as well—the key thing is that they are demonstrably different ideas. There are good reasons to believe that Nietzsche's perspectivism, like Descartes's skepticism, is also two-tiered. Nietzsche does present and promote perspectivism as a first-order theory or collection of philosophical theories, as discussed above (and as I defended in earlier work). But he also offers perspectivism as a second-order epistemological methodology.

Why would Nietzsche want methodological perspectivism? What is the point or advantage of it? The answer is to provide and enhance *understanding* — of the human condition, of society, morality, religion, science, music, politics. There are not only distinct perspectival truths, but the very manner in which these domains can be understood, and the systematic approaches we can take to provide that understanding, are diverse to the point of incommensurability. Nevertheless, our understanding is broadened precisely through the recognition of that diversity. The Uruguayan poet Eduardo Galeano observed in *Walking Words* that "The Church says: the body is a sin. Science says: the body is a machine. Advertising says: The body is a business. The Body says: I am a fiesta" (Galeano 1995, 151). Each of these is a legitimate perspective on the body, each can generate perfectly true claims within its investigative paradigm, and still each alone is incomplete, stunted, and provincial. It is only by appreciating all of these perspectives that a richer sort of understanding can be achieved. With first-order perspectivism Nietzsche is (partly) offering a theory of knowledge in which diverse perspectives generate distinct kinds of knowledge. Second-order perspectivism is a way of taking a stance on those first-order points of view and utilizing them to produce understanding.

2.5   Understanding

Understanding is a separate epistemic state from knowledge and has only recently received critical scrutiny. There are five ways in which epistemologists have distinguished understanding from knowledge. The first is *epistemic luck*. Jonathan Kvanvig has argued that while knowledge can be undermined by epistemic luck, understanding cannot be (Kvanvig 2017). For example, imagine someone who can correctly answer any question about the Comanche dominance of the southern plains in North America from 1775 to 1875. They have a good understanding of Comanche dominance even if they acquired this information in a way riddled with epistemic luck. They might have read a book on the Comanches that was filled with errors, but misremember the book in a way that corrects its mistakes. Such a scenario would typically be regarded as one in which the subject lacks knowledge due to the presence of epistemic luck, but it would still be a mistake to insist that they have no understanding of Comanche dominance.[4]

Some, notably Linda Zagzebski, have argued that "understanding, in contrast to [propositional knowledge] not only has internally accessible criteria, but is a state that is constituted by a state of conscious transparency. It may be possible to know without knowing one knows but it is impossible to understand without understanding one understands" (Zagzebski 2001, 246). The transparency thesis is the second claim often made on behalf of understanding.

Understanding is also supposed to be *valuable* in a way distinct from knowledge. Duncan Pritchard has argued that one might have knowledge without it being particularly connected to any kind of cognitive achievement (Pritchard et al. 2010, 80–84). A child who accepts a true belief about dinosaurs on the basis of parental testimony might have knowledge about dinosaurs, but that knowledge is not really creditable to the child's cognitive abilities. Pritchard thinks understanding is different, that it is a genuine cognitive achievement and is an epistemically internalist notion. Like Zagzebski, Pritchard holds that if one has understanding then it should not be opaque that one has such understanding. As a result, understanding is distinctively valuable. The value of knowledge runs the risk of being swamped by the value of merely possessing the truth, but the value of understanding consists in the virtue of achievement.

The transparency thesis of Zagzebski and Pritchard is dubious. People routinely do not know the things they think they know and can find out that they don't really believe the things they take themselves to believe. Studies in perceptual construction show that subjects routinely report seeing a light flash or a pinprick when told to expect one, even though there was no flash or prick. In short, our minds are far more opaque than we'd like to think, which makes it dubious that any particular

---

[4]Pritchard, Millar, and Haddock (2010, 77–78) challenged Kvanvig on this point, arguing that understanding is subject to some but not all forms of epistemic luck.

mental state like understanding should be perfectly transparent.[5] Certainly Nietzsche was doubtful about our continuing overestimation of consciousness (*GS* 11): "By far the greatest part of our spirit's activity remains unconscious and unfelt" (*GS* 333), and he calls "the absurd overestimation of consciousness" a "tremendous blunder" (*WP* 529). However, there are other common theses about understanding that are very Nietzschean indeed.

Knowledge is straightforwardly Boolean: either you know that *p* or you do not know that *p*. Understanding, on the other hand, comes in degrees. As Catherine Elgin writes, "A freshman has some understanding of the Athenian victory [at Marathon], while her teaching fellow has a greater understanding and her professor of military history has an even greater understanding" (Elgin 2017, 58). Or consider theories of planetary motion. Ptolemy had some understanding of the motion of the heavens — he realized that the sun, moon, stars, other planets, and Earth were in motion relative to each other, that the stars seemed not to move and that the sun and moon moved in a very regular pattern, whereas sometimes the planets seemed to loop back on their own orbits (retrograde motion) and that this needed explanation.

Copernicus had a better understanding, placing the sun stationary at the center of the system instead of the Earth, but his theory was just as false as Ptolemy's, since he still believed that orbits were circular. Kepler's astronomy provided a better understanding as he replaced Copernicus's circular orbits with elliptical ones. His flawed understanding was demonstrated by Newton, who showed that because the planets exert gravitational force on each other, orbits cannot be perfectly elliptical. Newton did not get the last word in either, as relativity theory establishes that there is no absolute space from which we can measure the motion of bodies. Instead of the Sun moving around the Earth, or the Earth moving around the sun, a better understanding recognizes that there is no absolute space or absolute motion in the way Newtonian mechanics suggested. The preceding is not a tale of one false theory being supplanted by another false theory, but of increasing degrees of understanding of a complex topic.

Perhaps the most distinctive and vital element of understanding is that false theories and erroneous models not only can lead to understanding, but understanding some phenomena may be impossible without them. In this way falsehood is an essential component of understanding the world. No one begins their scientific education by studying quantum mechanics and general relativity. Instead everyone starts with Newton's mechanics, his three laws of motion, and assumption of absolute space, even though all physicists reject Newton's physics and think that at best it is a special case of relativity physics. It would be a mistake to insist that studying Newton gives us no understanding of how the world works, even though his theory is, strictly speaking, false. Or take Euclidean geometry, which at most is true of ideal Platonic lines, points, and plane figures, but is not a true description of our imprecise and vaguely bounded material world. Not to mention the fact that Lobachevskian or Riemannian geometry more accurately describes spacetime.

---

[5] On the fallibility of conscious introspection, see Schwitzgebel (2011).

Nonetheless learning Euclidean geometry enhances one's understanding of area, volume, and angularity, and is essential to draftsmanship and the building trades.

Once you see how one can understand phenomena imperfectly, or by accepting imperfect theories, examples come up everywhere. Elgin offers Boyle's Gas Law, which states that the volume and pressure of a gas in a closed container vary inversely. She points out that Boyle's Law falsely treats gas molecules as dimensionless, frictionless, perfect spheres that exhibit no intermolecular attraction (Elgin 2017, 61). Still, recognizing the relationship between gas volume and pressure was an important bit of scientific progress. In philosophy the assumption of causal determinism advances debates in action theory even though no scientist thinks it is true of the quantum world. Or take the justified-true-belief analysis of knowledge — another false theory whose long-term acceptance advanced our understanding of many topics in epistemology.

A cartographic example of how false models can provide understanding is the London Underground. Compare the tangled mass of spaghetti strands that most accurately represents the topology of the tube lines to the more familiar, highly stylized map of the Underground.[6] A London resident or expert may prefer the first, more accurate map, but a first-time tourist will understand how to get around the city, grasp the general relationship of prominent locations, and navigate their way to their desired destination much better with the stylized map. It's no coincidence that it is the second map that is in all the tube stations and not the first. The spaghetti-strand map is similar to the problem of model overfitting in statistics. Imagine a scatterplot of data points. One can always find an equation that draws a curve through the data that matches it perfectly, but doing so is inferior to finding the curve or line that shows the general trend of the data. Figuring out the general trend allows us to better predict future observations than essentially memorizing each piece of past information.

Kvanvig (2017, 181) argues that understanding is at least quasi-factive — genuine understanding cannot be *too* far removed from the truth. In many cases he is right; a tube map with stations randomly scattered all over the city is no help at all, and a cosmology in which no part of the heavens moves at all gives us much less understanding of our experiences than Ptolemy. But our understanding is much improved with models that are not overfit, stylized maps, and (except in extremely specialized cases) Euclidean geometry.

The mainstream literature on understanding treats it as a matter of degree and resulting from idealized models. It bears mentioning that there are recent criticisms of allowing idealizations (which perforce include falsehoods) to provide any sort of epistemic value. Sullivan and Khalifa (2019), for example, argue that if idealizations produce any epistemic value at all, it is always inferior to that yielded by more accurate models. As a result, there is no motivation to use them when a better model is available. More promising, they suggest, is a treatment of understanding that is non-epistemic, and that the proper use of idealizations is had "by flagging

---

[6]Accurate map: http://content.tfl.gov.uk/london-connections-map.pdf. The familiar stylized map: http://content.tfl.gov.uk/standard-tube-map.pdf

irrelevancies, explaining, structuring contrastive explanations, isolating causes, and imparting modal information", all tasks they regard as pragmatically useful, but not of epistemic value. A deep dive into these waters is beyond the scope of this paper (and Khalifa and Millson address perspectivism and truth in Chap. 6 in this collection). It is enough here to give a novel interpretation of Nietzsche as offering a rudimentary perspectival treatment of understanding that anticipates some of the contemporary discussion in epistemology and philosophy of science.

The idea that methodological perspectivism aims to enhance our understanding gives a new way to interpret Nietzsche's claims about untruths being a condition of life and the epistemic value of errors. In fact, it predicts that he would say such things. If when Nietzsche discusses knowledge he is at least sometimes fumbling towards the idea of understanding, then what he says about the usefulness of errors snaps neatly into place. "Delusion and error are conditions of human knowledge and sensation", Nietzsche writes (*GS* 107), "we simply lack any organ for knowledge, for 'truth': we 'know' (or believe or imagine) just as much as may be *useful* in the interests of the human herd, the species…" (*GS* 354). In *WP* 503 he writes, "The entire apparatus of knowledge is an apparatus for abstraction and simplification". Perhaps the best known passage along these lines is *BGE* 4, in which Nietzsche writes,

> the falseness of a judgment is for us not necessarily an objection to a judgment …. The question is to what extent it is life-promoting, life-preserving, species-preserving, perhaps even species-cultivating … without accepting the fictions of logic, without measuring reality against the purely invented world of the unconditional and self-identical, without a constant falsification of the world by means of numbers, man could not live — that renouncing false judgments would mean renouncing life and a denial of life. To recognize untruth as a condition of life … a philosophy that risks this would by that token alone place itself beyond good and evil.

Those ideas sound shockingly radical, unless you read "knowledge" as "understanding" and interpret his endorsement of fictions and falsehoods as essential to life as an acknowledgment of the epistemic value of idealized models. Seen through that lens, the passages just cited become perfectly sensible.

For all the reasons discussed above, our understanding of the world is improved by false models, and we understand just as much as may be useful to us. Nietzsche argues in *GS* 307 that when we give up former beliefs we mistakenly chalk it up to a victory for reason. Rather, it was an opinion useful for our former selves that we no longer need, like a snake shedding a skin it has outgrown. So the understanding provided by Newtonian physics to a novice student is outgrown by the mature scientist, and the navigational understanding that the London tube map provides to a tourist is ultimately surpassed when the tourist becomes a long-time resident. What we really want is "to schematize — to impose upon chaos as much regularity and form as our practical needs require" (*WP* 515); that is how we make the world intelligible and useful to us. In an early fragment of a critique of Schopenhauer, Nietzsche wrote that "The errors of great men are venerable because they are more fruitful than the truths of little men."[7] Why are they more fruitful? Because those great errors – the erroneous yet magnificent edifices of Ptolemy, Galileo, Newton, Aristotle – all contributed to how we understand the world far more than the tedious minor truths worked out by their followers.

Nietzsche's second-order epistemological methodology is that understanding is improved through the adoption of different perspectives, like recognizing that polar coordinates are trigonometric transforms of Cartesian coordinates. "Physics too is only an interpretation ... and *not* a world-explanation", he famously writes in *BGE* 14. How can physics, our most fundamental and successful science, just be an *interpretation*? But if instead *BGE* 14 is really a statement of methodological perspectivism, then it sounds downright reasonable. Physics is one tool to understand the world, one perspective of great reach and fecundity. At the same time it would be foolish to expect physics to help us understand the aesthetic dimensions of *Starry Night* or *Mass in D-Minor*. Nor are we going to settle the relative merits of deontology and consequentialism using the mathematical language of final physics. Physics isn't a world-explanation because there are so many parts of reality — art, music, love, meaning, virtue — beyond its purview. Nietzsche is not denying that science produces knowledge. He is denying that it is the lone method by which we should understand reality (*GS* 373–374).

The idea that understanding a topic is improved by multiple standpoints is given additional support by Nietzsche's own practice. His literary style is the most diverse of any philosopher: with aphorisms, extended essays, an intellectual autobiography, a quasi-religious book with characters and action, poetry, and critical papers. All are examples of perspectival practice, addressing problems and ideas from the greatest multiplicity of approaches. If an analytic argument does not enlighten us, maybe the Songs of Prince Vogelfrei will. If brief aphorisms aren't helping us figure things out, maybe a long parable will. Couldn't make heads or tails of what he's after in the aphoristic *Beyond Good and Evil*? Let's try the nearly analytic essays of *On the Genealogy of Morals*. The great variety of tones, or voices of style, also lend support. Nietzsche can be elegiac, scornful, bombastic, ironic, witty, and seldom engages in the plodding self-seriousness of most philosophic writing. His pyrotechnics are the very opposite of Kant, whose writing Nietzsche described as something to be endured, the result of a "deformed concept-cripple" who cannot make his words dance (*TI*, "What the Germans Lack" §7). His verbal fireworks are meant to do more than entertain, though, each is intended in its own way to illuminate.

Nietzsche's first order perspectivism is a genuine part of his thought, and something he seriously advances. But it should be seen as a component of his larger architectonic, his kaleidoscope of voices, approaches, and arguments, all aiming at the greatest understanding possible. Nietzsche did not make the distinction between knowledge and understanding, and the extent to which he muddled these concepts together helps explain the difficulty in making sense of his epistemic perspectivism.

---

[7] Cited in Kaufmann (1982, 30).

Once that difference is kept firmly in mind, first-order perspectivism and second-order methodological perspectivism naturally divide and help make sense of a great deal of Nietzsche's epistemological ruminations.

Acknowledgements For helpful comments on earlier versions, I am very grateful to (1) Michela Massimi and the participants in the Knowledge From a Human Point of View conference at the University of Edinburgh, especially Dave Ward, (2) Manuel Liz, Marga Vásquez, and the participants in the Points of View workshop at the Universidad de la Laguna in Tenerife, and (3) the audience at the 24th World Congress of Philosophy in Beijing. Work on this paper was supported by Research Project FFI2014-57409, *Points of View, Dispositions and Time. Perspectives in a World of Dispositions* (Spain).

Bibliography

Alston, W. P. (1991). *Perceiving God: The epistemology of religious experience*. Ithaca: Cornell University Press.
Babich, B. (1994). *Nietzsche's philosophy of science*. Albany: State University of New York Press.
Berry, J. N. (2011). *Nietzsche and the ancient skeptical tradition*. Oxford: Oxford University Press.
Danto, A. (1965). *Nietzsche as Philosopher*. New York: Macmillan.
Doyle, T. (2009). *Nietzsche on epistemology and metaphysics: The world in view*. Edinburgh: Edinburgh University Press.
Elgin, C. Z. (2017). *True enough*. Cambridge: MIT Press.
Galeano, E. (1995). *Walking words*. New York: W. W. Norton.
Granier, J. (1977). Perspectivism and interpretation. In D. B. Allison (Ed.), *The New Nietzsche* (pp. 190–200). New York: Dell.
Grimm, R. H. (1977). *Nietzsche's theory of knowledge*. Berlin: Walter de Gruyter.
Hales, S. D. (forthcoming). Self-refutation. In *The Routledge handbook of relativism*. Martin Kusch/London: Routledge.
Hales, S. D., & Welshon, R. (2000). *Nietzsche's Perspectivism*. Urbana-Champaign: University of Illinois Press.
Hollingdale, R. J. (1973). *Nietzsche*. London: Routledge and Kegan Paul.
Kaufmann, W. (Ed.). (1982). *The Portable Nietzsche*. New York: Penguin.
Kvanvig, J. L. (2017). Understanding. In F. D. Aquino & W. J. Abraham (Eds.), *Oxford handbook on the epistemology of theology* (pp. 175–190). Oxford: Oxford University Press.
Lightbody, B. (2010). Nietzsche, Perspectivism, anti-realism: An inconsistent triad. *The European Legacy, 15*(4), 425–438.
Mittelman, W. (1984). Perspectivism, becoming, and truth in Nietzsche. *International Studies in Philosophy, 16*(2), 3–22.
Poellner, P. (1995). *Nietzsche and metaphysics*. Oxford: Oxford University Press.
Pritchard, D., Millar, A., & Haddock, A. (2010). *The nature and value of knowledge*. Oxford: Oxford University Press.
Reginster, B. (2001). The paradox of Perspectivism. *Philosophy and Phenomenological Research, 62*(1), 217–233.
Schrift, A. (1990). *Nietzsche and the question of interpretation*. New York: Routledge.
Schwitzgebel, E. (2011). *Perplexities of consciousness*. Cambridge: MIT Press.
Sinhababu, N. (2017). Nietzschean Pragmatism. *The Journal of Nietzsche Studies, 48*(1), 56–70.
Sullivan, E, & Khalifa, K. (2019). Idealizations and understanding: Much ado about nothing? *Australasian Journal of Philosophy* forthcoming.
Zagzebski, L. (2001). Recovering understanding. In M. Steup (Ed.), *Knowledge, truth, and duty: Essays on epistemic justification, responsibility, and virtue* (pp. 235–252). Oxford: Oxford University Press.

Chapter 3
Pluralism and Perspectivism in the American Pragmatist Tradition

Matthew J. Brown

Abstract This chapter explores perspectivism in the American Pragmatist tradition. On the one hand, the thematization of perspectivism in contemporary epistemology and philosophy of science can benefit from resources in the American Pragmatist philosophical tradition. On the other hand, the Pragmatists have interesting and innovative, pluralistic views that can be illuminated through the lens of perspectivism. I pursue this inquiry primarily through examining relevant sources from the Pragmatist tradition. I will illustrate productive engagements between pragmatism and perspectivism in three areas: in the pragmatists' fallibilistic theories of inquiry and truth, in their pluralistic metaphysics, and in their views on cultural pluralism. While there are some potential sticking points between pragmatism and perspectivism, particularly around the visual metaphor of perspective, these philosophical approaches nonetheless have much to learn from each other. Perspectivism is in danger of falling between the horns of pernicious relativism and a platitudinous view of the limits of human perception and cognition. The pragmatists accounts of truth and reality open the possibility of a more thoroughgoing perspectivism. I will follow this thread through Charles S. Peirce's, William James', and John Dewey's theories of inquiry and truth, Peirce's evolutionary metaphysics, James' radical pluralism, Dewey's cultural naturalism, Richard Rorty's anti-essentialism, Jane Addams' standpoint epistemology, W.E.B. Du Bois' theory of race consciousness, Horace Kallen's and Alain LeRoy Locke's cultural pluralism, and Mary Parker Follett's account of pluralistic integration.

Keywords American pragmatism · Perspectivism · Fallibilism · Inquiry · Truth · Standpoint epistemology · Race consciousness · Double consciousness · Cultural pluralism · Metaphysical pluralism · Integrative pluralism · Charles Peirce · William James · John Dewey · Richard Rorty · Jane Addams · W.E.B. Du Bois · Alain Locke · Mary Parker Follett · Horace Kallen

M. J. Brown (✉)
University of Texas at Dallas, Richardson, TX, USA
e-mail: mattbrown@utdallas.edu

© The Author(s) 2020
A. Crețu, M. Massimi (eds.), *Knowledge from a Human Point of View*,
Synthese Library 416, https://doi.org/10.1007/978-3-030-27041-4_3

3.1   Introduction

The American pragmatist tradition, the central movement of the American philosophical tradition from the late nineteenth until the mid-twentieth century, is a diverse and complex philosophical tradition independent from, though in dialogue with, the dominant, so-called 'analytic' and 'continental' philosophical traditions. One core commitment of nearly all the pragmatists is to conceive of knowledge 'from a human point of view', and to see the ramifications of that epistemic stance also for our ethical, political, and metaphysical views. In this paper, I will trace a variety of perspectivist themes, or philosophical ideas that perspectivism can illuminate, through pragmatists from Charles Peirce to Richard Rorty, taking a broad and inclusive view of the membership of that tradition. I will attempt to draw out paradigmatic pragmatist views and what they tell us about perspectivism and pluralism.

Perspectivism involves a problematic metaphor from the point of view of the pragmatist tradition. At heart, perspectivism provides a *visual* metaphor, which in turn suggests two ideas that the pragmatists took great pains to refute and replace. The first idea is *the spectator theory of knowledge*, according to which vision (traditionally understood) is the best metaphor for knowledge in general. The knower is understood to be a passive receiver of information about the known, as the spectator passively receives visual impressions of that which is viewed. The second idea is the dichotomy between appearance and reality, which posits general, philosophically significant distinctions between the real and the merely apparent. Perspectivism suggests a situation in which we each 'stand' in a different place, with a partial view of one and the same 'real' object, receiving partial and limited knowledge about it *from our point of view*.

I shall discuss the difficulties involved below in much greater detail. But there remain a variety of resonances between the American pragmatist tradition and contemporary perspectivist philosophy: a commitment to some form of pluralism and to the recognition of the limits of human knowledge that could be described alternatively as anti-absolutism, anti-objectivism, or anti-fundamentalism. Furthermore, the pragmatists have valuable resources to offer the perspectivists, including a simpatico metaphilosophical orientation, a pluralistic and anti-reductionist metaphysics, a sophisticated contextualism and fallibilism, and non-dominationist and pluralistic ideas about building bridges and relations of reciprocity between diverse perspectives.

Pragmatism is best known, perhaps, as a suite of connected views on the nature of belief, meaning, inquiry, and truth, and especially the latter. Most philosophers will have at least encounterd a certain cartoon version of the pragmatist theory of truth, on the basis of which they widely but entirely mistakenly believe it to be an untenable approach to truth. Some are likely to know the pragmatist theory of meaning, also known sometimes as "the pragmatic maxim", according to which the meaning of a concept or claim is to be elucidated by how it plays out in *practice*, its practical connections and implications, broadly construed. The pragmatist views of belief and inquiry that some philosophers, especially philosophers of science, might be familiar with are part of a powerful fallibilist, contextualist epistemology that attends to the roles of values or purposes in knowledge. These views are reflections of, or perhaps generalize to, a metaphilosophy: philosophical questions are to be answered not *sub specie aeternitatis*, but by looking to the role they play in what is variously called practice, experience, life, or culture. That is, philosophical questions can only be answered *from a human point of view*.

Other, less well-known philosophical views relevant to perspectivism and to philosophy from a human point of view also occupy an important place in the American pragmatist tradition. Pragmatists have tended to be lumped among other anti-metaphysical philosophers of their time, but many pragmatists have their own positive, interesting metaphysical views that come out of their broader pragmatist metaphilosophy. These views reinforce the pragmatists' fallibilist epistemology and their views on truth. Even less well known are the ways in which the pragmatists were sensitive not only to the plurality of human experiences, practices, and cultures, but the sensitivity of some of them to issues of power intersection of those differences, and their commitment to non-dominationist, reciprocal encounters across differences.

The pragmatists are far from univocal; there is no single pragmatist philosophical system. On substantive issues in all of these areas, major pragmatist thinkers have disagreed. Rather than oversell the amount of consensus among the American pragmatist tradition, I will from here on treat their diverse but interconnected body of thought as a toolbox from which the contemporary perspectivist might find several tools for thinking about knowledge from a human point of view. I will organize my discussion into three parts: epistemological views about inquiry and truth, views about the metaphysical background, and accounts of cultural diversity, pluralism, and integration.

3.2   The Pragmatists' Fallibilistic Theories of Inquiry and Truth

One of the most well-known elements of the classical pragmatist philosophy is the theory of inquiry. From Charles Peirce's doubt-belief scheme in his writings on "the logic of science", to William James's discussion of "the will to believe", to John Dewey's writings on education, intelligence, and logic, the account of belief-formation and inquiry is a key element of the tradition, though each writer emphasizes different elements. Each focusses on different ways in which knowledge is constructed from a human point of view. Peirce is particularly interested in the way that science represents a *communal* way of settling belief, while James wants to also accommodate personal belief, including religious belief. Dewey's theory of inquiry attempts to take the larger bio-cultural environment into account in order to provide a contextual theory of inquiry.

Each of these thinkers attempted to carefully tie truth to their theories of inquiry. Rather than thinking of truth as a semantic or metaphysical notion, the pragmatists attempt to analyze the role of truth in our practices of inquiry. This is not the same as providing an epistemic theory of truth like verificationism or ideal assertibility. As Cheryl Misak has argued, the pragmatists instead provide a 'pragmatic elucidation' of the concept of truth, i.e., an explanation of the role that the concept plays in our practices and lived experiences, rather than an analytic definition or criterion of truth (Misak 2004). The pragmatists had no particular quarrel with correspondence as an analytic definition of truth; they merely saw it as formal and empty, revealing little of our uses of the concept, and as tending to lead to bad metaphysical dualisms (Capps 2019).

3.2.1   Charles Peirce's Doubt-Belief-Inquiry Schema

Charles Sanders Peirce's account of inquiry focuses on the fixation of belief in response to doubt. Peirce and the other pragmatists follow Alexander Bain in defining beliefs as "habits of action" (Bain 1859; Fisch 1954; Haack 1982). Doubt is the result either of the thwarting of a belief by experience, the lack of a belief providing a habitual response to a situation, or the conflict between one's belief and the beliefs of others; it results in hesitancy and irritation. Inquiry is the overcoming of doubt by forming a new belief. Where no real and living doubt exists, on the other hand, no inquiry is possible—an argument that Peirce deployed against Descartes.

For Peirce, convergence towards what he called the "final opinion" on any specific question is a regulative ideal of inquiry, and we understand *truth* in terms of that final opinion. In the most well-known formulation, "[t]he opinion which is fated to be ultimately agreed to by all who investigate, is what we mean by truth, and the object of this opinion is the real" (Houser and Kloesel 1992, 1:139). That is, a belief is true if it is "unassailable by doubt" (Peirce Edition Project 1998, 2:336). This is not a prophecy about the future of inquiry, but a counterfactual claim: whatever view would be the considered belief of indefinitely extended inquiries is what we ultimately mean by 'the truth of the matter'.

Nonetheless, there is room for pluralism and a kind of perspectivism in Peirce's view. Peirce's accounts of doubt, scientific method, and convergence all depend on the idea of a *community of inquiry*. "We individually cannot reasonably hope to attain the ultimate philosophy which we pursue; we can only seek it, therefore, for the *community* of philosophers" (Houser and Kloesel 1992, 29). And again:

> The real, then, is that which, sooner or later, information and reasoning would finally result in, and which is therefore independent of the vagaries of me and you. Thus, the very origin of the conception of reality shows that this conception essentially involves the notion of a COMMUNITY, without definite limits, and capable of an indefinite increase of knowledge. And so those two series of cognitions—the real and the unreal—consist of those which, at a time sufficiently future, the community will always continue to reaffirm; and of those which, under the same conditions, will ever after be denied. Now, a proposition whose

falsity can never be discovered, and the error of which therefore is absolutely incognizable, contains, upon our principle, absolutely no error. Consequently, that which is thought in these cognitions is the real, as it really is. There is nothing, then, to prevent our knowing outward things as they really are, and it is most likely that we do thus know them in number-less cases, although we can never be absolutely certain of doing so in any special case (Houser and Kloesel 1992, 1:52, capitalization in original.)

This communal notion is important in part because it is the beliefs of others differ-ing from your own that provide one of the positive reasons to doubt that occasions inquiry in the first place. If in the long run, inquiry should lead to convergence, on the way there, it depends on inquirers coming at problems from different perspectives.

3.2.2 William James's Liberalization of Peirce

James did not differ in the basics from Peirce. He emphasized two points already nascent in Peirce's account: the purposive nature of human belief and the contin-gency of present belief as compared to the destined or final opinion. The difference of emphasis within broadly shared ideas about belief and inquiry led James to a more pluralistic and permissive theory of knowledge and to a quite different theory of truth.

James combined the view that beliefs were habits of action, as Peirce had held, with the recognition, drawn from his psychological work, that humans have many purposes for which they act. It follows from this combination that, for James, how inquiry proceeds to settle belief would concern human purposes. James furthermore emphasized the contingency of belief. We see with Peirce already the idea that one's current belief depends on the range of experiences and interactions one has had that would cause one to doubt it. In the absence of a positive reason to doubt, belief for Peirce is settled; what one is prepared to believe and to doubt is thus conditioned by one's history and experience. James expands this sort of contextualism to the great variety of human purposes beyond the narrowly scientific.

These views of James lead to his famous argument in *The Will to Believe* (James 1896). There James considers the tension between two epistemic "laws"— "Believe truth! Shun error!". That is, he considers the trade-off between false posi-tive and false negative errors (Magnus 2013). In cases where we face options between what to believe, the options are genuine ones, believable based on what we know and our existing epistemic commitments, where we cannot put off the ques-tion indefinitely, and we do not have sufficient evidence to decide the question, we must decide what to believe based on our "passional nature". One natural interpreta-tion is that we must decide based on what our purposes or our values tell us about the trade off between the two types of errors, an early statement of the argument from inductive risk (see Magnus 2013 on the "James-Rudner-Douglas thesis"). When different people have different goals and values, we should expect in this situ-ation that they will come to believe different things.

Contrary to Peirce's focus on convergence, James thus emphasized epistemic pluralism. And at least on one place, James articulates a form of pluralistic tolerance highly in tune with more recent pluralist and perspectivist ideas—"Hands off: neither the whole of truth, nor the whole of good, is revealed to any single observer, although each observer gains a partial superiority of insight from the peculiar position in which he stands" (James 1899).

James was more liberal than Peirce in applying the term "truth". He wanted to recognize not only the opinion that would prove unassailable in the long run (about which we may have little use), but also those beliefs that had proved particularly successful in more immediate contexts as being "true" in some sense. He based his thinking on truth on the gap between present need and the imagined future opinion of Peirce, distinguishing "temporary truth" in his sense from "absolute truth" in Peirce's sense. James sometimes speaks of truth as what is expedient, or useful, or good to believe for definable reasons. In another passage, he writes about truth:

> It means, they say, nothing but this, *that ideas (which themselves are but parts of our experience) become true just in so far as they help us to get into satisfactory relation with other parts of our experience*, to summarize them and get about among them by conceptual shortcuts instead of following the interminable succession of particular phenomena. Any idea upon which we can ride, so to speak; any idea that will carry us prosperously from any one part of our experience to any other part, linking things satisfactorily, working securely, simplifying, saving labor; is true for just so much, true in so far forth, true *instrumentally* (James 1907, 58).

And elsewhere in the same text, James claims that, "[t]rue ideas are those that we can assimilate, validate, corroborate and verify" (ibid., 201). Here, not just belief but truth itself is dependent upon the activities (and thus the purposes) of human actors.

3.2.3 John Dewey's Situational Theory of Inquiry

John Dewey took this view of things further. Dewey embedded Peirce's and James's conception of belief within a biological and psychological picture of an active creature navigating an uncertain and changing world. The original need for inquiry derives from the need for the creature to respond to situations where it is in disequilibrium with its environment, to remake both its habits and its environment such that the creature could draw support from the environment. To this, he added the point that the environment for human inquirers is cultural as well as physical. Inquiry thus transforms not only (or primarily) individual beliefs and habits but also cultural representations, tools, practices, and institutions. This is the position Dewey refers to in *Logic: The Theory of Inquiry* as "*cultural naturalism*" (Dewey 1938, 12:28).

These additions create a richer, but also more contextualist account of inquiry. Dewey highlighted this by describing *inquiry* not as the fixation of belief but as the settling of a "problematic" or "indeterminate *situation*", what Dewey also called, following Jane Addams, a "perplexity" (Addams 1902; Dewey 1933). Situations become indeterminate or problematic when the activities of the organisms or actors fail to function as expected; habits and values in that case no longer guide the activity coherently. The situation is resolved by transforming it so that it is more "unified", i.e., so that the inquirer and their natural and cultural environment interacts in such a way that is no longer problematic. Dewey here replaces the individualistic, psychological language of "doubt" and "belief" with an account that is both more general (allowing, for example, for group or social inquiry) and more ecological, decentering the mind of the believer.

As inquiry is directed at resolving situations, its results are likewise situational. From the perspective of one group of inquirers in a particular natural and cultural environment, given certain practices and aims, in response to particular problems that arise, one judgment may be correct; from the perspective of differently-situated inquirers, it may not be. At least, further inquiry in the new situation would be required to determine whether it was.

Dewey was more wary of using the term "truth", especially after uncharitable interlocutors like Bertrand Russell persistently misinterpreted what he had to say about the matter. If Dewey had a theory of truth, or at least a pragmatic *elucidation* of the concept of truth (see Capps 2018; Misak 2004), it is this: to call a judgment "true" is just to say in retrospect that it successfully resolved the problematic situation that the inquiry that produced it was occasioned by. This account fits with Dewey's quite enigmatic statement about truth:

> In contrast with this view, my own view takes correspondence in the operational sense it bears in all cases except the unique epistemological case of an alleged relation between a 'subject' and an 'object'; the meaning, namely, of *answering*, as a key answers to conditions imposed by a lock, or as two correspondents "answer" each other; or, in general, as a reply is an adequate answer to a question or a criticism—as, in short, a *solution* answers the requirements of a *problem*...
>
> In the sense of correspondence as operational and behavioral (the meaning which has definite parallels in ordinary experience), I hold that my *type* of theory is the only one entitled to be called a correspondence theory of truth (Dewey 1941, 178–79).

Dewey's account of truth is, like James', sensitive to human values and purposes, and even more thoroughly contextual than James'.

3.2.4 Epistemological Lessons for and from Perspectivism

According to these pragmatists, inquiry, belief-formation, or knowing are responses to doubts or problems that arise in the course of human practices and activities. Not only are they thoroughly grounded in a human point of view, they are grounded in human need, experience, values, and culture. Not only is belief or knowledge in some sense contextual or perspectival, but so is truth itself, at least for James and Dewey. The context is an active one belied by visual metaphors for knowledge; belief, knowledge, and truth have as much to do with how we make and re-make the world as with what we find there when we look.

These pragmatist theories of inquiry and truth are thoroughly fallibilistic. According to Hilary Putnam, "[t]hat one can be both fallibilistic *and* antiskeptical is the basic insight of American pragmatism" (Putnam 1992, 29). Fallibilism suggests that knowledge is incomplete or revisable, that what we take as truth may be partial or replaced entirely. In the case of Peirce, James, and Dewey, fallibilism is defended not merely as a prudent attitude in the face of the limits of human knowledge, but also as supported by a basic metaphysical worldview. It is thus to the much less well-known and well-appreciated pragmatist metaphysical views that I now turn.

3.3 Pluralism and Perspectivism in Pragmatist Metaphysics

The pragmatists did not see a need to *ground* their fallibilist, contextualist epistemology in a metaphysical picture, nor vice versa. Rather, they saw their basic metaphilosophical orientation as that of seeking the answer to philosophical questions in human practice, as leading to new insights in metaphysics. These insights in turn undermined many of the critiques of pragmatist epistemology deriving from varieties of idealist and realist commitments. Indeed, they were able to situate the pragmatists' epistemic insights in a broader context, and so, despite not being tightly coupled in a philosophical "system", their metaphysical and epistemological approaches tended to support one another.

The pragmatists' metaphysical views are in many ways unorthodox by contemporary lights, but they provide potential avenues for perspectivists and others focused on working out a theory of knowledge from a human point of view for escaping certain metaphysically-grounded criticisms. They provide an alternative picture to the dualistic metaphysics often erroneously drawn out of thinking about the definition of truth and representationalist theories of knowledge.

3.3.1 Peirce's Triadism and Evolutionary Metaphysics

Peirce's philosophical works are full of three-way distinctions. We've already seen one: belief, doubt, and inquiry. His logic described three fundamental forms of inference: deduction, induction, and abduction. His version of pragmatism is presented as a third degree of clarity, i.e., a third way of clarifying the meaning of an idea: to tacit familiarity and abstract definition he added pragmatic clarification—tracing the consequences of an idea or concept for our habits of conduct. "Consider what effects, which might conceivably have practical bearings, we conceive the object of our conception to have. Then, our conception of these effects is the whole of our conception of the object" (Houser and Kloesel 1992, 1:32). Could it just be a coincidence, or a quirk of Peirce's psychology, that he saw things in terms of threes?

No, indeed, Peirce saw three universal categories—"categories" in the Aristotelian or Kantian sense—underlying all human knowledge and reality itself. The three categories in the abstract are named *firstness*, *secondness*, and *thirdness*, and one simple way to think of them is in connection with monadic, dyadic, and triadic predicates in logic.[1] *Firstness* concerns being or quality; *secondness* concerns relation or reaction; *thirdness* concerns mediation.

The centrality of triads to Peirce's philosophy means that *mediation* is central to Peirce's philosophical analyses. For instance, consider Peirce's semiotics (theory of signs). The meaning of a sign is not reducible, for Peirce, to a dyadic relation between a sign and the object it refers to. This relation is mediated by what Peirce calls the *"interpretant"*, which can be understood in terms of the effect the sign has on the one who interprets it, their understanding of or translation of the sign. This makes Peirce's theory of meaning irreducibly *perspectival*, dependent upon the interpreter of the sign. Peirce understands thought, language, mind, knowledge, and even the metaphysical nature of reality in terms of these mediational, often perspectival, processes.

Peirce's speculative metaphysics is also founded on a triad. First, there is pure chance or spontaneity, then there is mechanical determinism, and the mediating third is what Peirce calls "habit-taking". In his view, our experience and our science justify belief in the existence of real chances, a doctrine he called *"tychism"*. But there were also phenomena that showed more or less mechanical orderliness. While he held that determinism was false (and radically, for someone living prior to the development of quantum physics, he rejected determinism on *scientific* grounds), he did not believe in complete disorder. The mediating factor, for Peirce, was the tendency of stuff to become more orderly, to take on habits. Peirce thought the fundamental stuff of the universe was mind-like in nature, precisely in its tendency to form intelligible patterns. Matter, for Peirce, was just "effete mind", mind-stuff that had become so fixed in its habits as to largely lack spontaneity.

Peirce's cosmology involved an evolution of the universe from a state of pure chance or spontaneity, though the process of habit-taking, towards an end-state of pure order, a view Peirce called *"agapism"* (the process of habit-taking, for Peirce, being connected with a general principle of growth and with *agape* or self-sacrificing love). Given that we live somewhere in the middle of this process, we are never justified in believing that we have converged on the ultimate truth or in treating any fact as ultimate (McKenna and Pratt 2015, 66). Ours is a universe where fallibilism is the necessary attitude.

---

[1] Peirce held that higher-order n-place predicates were reducible to the first three, but that the first three were irreducible; he provided a formal argument to this effect.

3.3.2 James's Pluralistic Universe

William James's metaphysical writings are based in two fundamental commitments: *radical empiricism* and *melioristic pluralism*. Radical empiricism has three faces: methodological, psychological, and ontological. Methodologically, radical empiricism is a thesis about what we must regard as real: "Everything real must be experienceable somewhere, and every kind of thing experienced must somewhere be real" (James and Perry 1912, 3:81). As with classical empiricism, James *excludes* metaphysical posits that cannot be founded in experience; more radically, he suggests that every kind of experience must be *included* in our metaphysical picture. This becomes truly radical in his discussions of religious experience and divine existence.

Psychologically, radical empiricism is a thesis about the contents of experience; contra classical empiricism, James argues that experience is not atomistic. That is, we directly experience not only individual sensory qualities but also *connections* and *relations* between those qualities. Causal and logical relations, tendencies and processes of change are found in experience, not posited to explain the succession of separate experiential qualities. They are directly felt as part of what James calls "The Stream of Thought" (James 1890) or "The Stream of Consciousness" (James 1892). The metaphor of a stream is meant by James to indicate the fundamental continuity of experience, as opposed to the Lockean-Humean atomistic account.

Radical empiricism, in its final form, also involves a metaphysical thesis. James takes "pure experience" as the basic *stuff* out of which reality is made. "Experience" here is playing a different role than in psychology, where it is something had by an individual mind. Here, experience is taken as metaphysically basic, and whether some experience is qualified as matter or mind is more a matter of the relations it bears to other experiences than some inherent property in it. In "How Two Minds Can Know One Thing", James argues that the same experience (e.g., of one physical object) can be known to two different people by coming into relation to their different conscious experiences (James and Perry 1912, 61ff). We can see this as a kind of non-representationalist perspectival knowledge.

The metaphysical thesis of radical experience has sometimes been awkwardly labeled "neutral monism", and it has also been understood as a kind of panpsychism. But we must square this with the fact that James frequently wrote in opposition to what he called "monism" and defended a view he called "pluralism". For James, monism was the idea that everything was connected and subsumed into a single whole. The prominent form of monism in James's day was absolute idealism, deriving from Hegel. According to James, monism did not respect the reality of finite human experience, could not explain the evil and irrationality of the world, and was fatalistic or deterministic in a way that denied the reality of human freedom.

James's pluralism was the negation of monism, the idea that there is no all-encompassing, top-down whole. According to James, the particular is more basic than the general, and there are real disconnections, tensions, fluxes, incommensurabilities, novelties, and spontaneities in the world. Of course, James's pluralism is not absolute: there are real connections, but there are also experiences that lack any definite connections. What's more, James's pluralism is melioristic in that new connections can always be made. Echoing Peirce, James says, "[a]nd finally it is becoming more and more unified by those systems of connexion at least which human energy keeps framing as time goes on" (James 1907, 156).

3.3.3 John Dewey's Immediate Empiricism and Cultural Naturalism

Similar to James's radical empiricism, John Dewey defends a view that he calls *'immediate empiricism'*. According to Dewey, "[i]mmediate empiricism postulates that things—anything, everything, in the ordinary or non-technical use of the term 'thing'—are what they are experienced as" (Dewey 1910, 227). This account of reality is inherently perspectival. Dewey continues,

> Hence, if one wishes to describe anything truly, his task is to tell what it is experienced as being. If it is a horse that is to be described, or the *equus* that is to be defined, then must the horse-trader, or the jockey, or the timid family man who wants a 'safe driver', or the zoologist or the paleontologist tell us what the horse is which is experienced. If these accounts turn out different in some respects, as well as congruous in others, this is no reason for assuming the content of one to be exclusively 'real', and that of others to be 'phenomenal'; for each account of what is experienced will manifest that it is the account *of* the horse-dealer, or *of* the zoologist, and hence will give the conditions requisite for understanding the differences as well as the agreements of the various accounts. And the principle varies not a whit if we bring in the psychologist's horse, the logician's horse or the metaphysician's horse (Dewey 1910, 393-94).

One and the same horse may be experienced as, from different perspectives, a mode of transportation, an item of commerce, a beloved pet, a thing of beauty, or a biological specimen.

Dewey is at pains to distinguish his point from the claim that things are all and only what they are *known* to be. He rejects any simple equation of experience with knowledge, "[f]or this leaves out of account what the knowledge standpoint is itself *experienced as*" (Dewey 1910, 229–30). That is, knowing is a particular kind of experience, but not one that is closer to things as they *really* are than any other experience. Dewey everywhere resists sorting experiences into the really-real and mere appearance:

> … the chief characteristic trait of the pragmatic notion of reality is precisely that no theory of Reality in general, *überhaupt*, is possible or needed. It occupies the position of an emancipated empiricism or a thoroughgoing naïve realism. It finds that 'reality' is a *denotative* term, a word used to designate indifferently everything that happens. Lies, dreams, insanities, deceptions, myths, theories are all of them just the events which they specifically are. Pragmatism is content to take its stand with science; for science finds all such events to be subject-matter of description and inquiry—just like stars and fossils, mosquitoes and malaria, circulation and vision. It also takes its stand with daily life, which finds that such things really have to be reckoned with as they occur interwoven in the texture of events (Dewey 1917, 55).

Dewey's view is not a form of subjective relativism or even of the panpsychism that James seems to be suggesting. Dewey does not identify experience as something *subjective* or even something purely *mental*. The horse is part of our experience because (and insofar as) we physically interact with the horse when we experience it, by looking at it, riding it, brushing it out. The ways we experience it are modes of activity. About the term experience, Dewey says,

> Its nearest equivalents are such words as 'life', 'history', 'culture' (in its anthropological use). It does not mean processes and modes of experiencing apart from *what* is experienced and lived. The philosophical value of the term is to provide a way of referring to the unity or totality between what is experienced and the way it is experienced... (Dewey 1922, 351; see Alexander 2014)

This is the broader metaphysical version of Dewey's *cultural naturalism*. The recognition here is that human experience encompasses everything that the anthropologist might refer to as "cultural practices".

3.3.4   Richard Rorty's Anti-Essentialism

In "A World without Substances or Essences" (1999), Richard Rorty describes his form of pragmatism as a type of *anti-essentialist panrelationalism*. On Rorty's view, nothing has an essence or essential properties; everything is constituted by its relations to other things, and there is no way to demarcate intrinsic from extrinsic relations. He motivates this view as the way of thinking shared by many Anglophone and non-Anglophone philosophers, despite the so-called 'analytic'-'Continental' split:

> The quickest way of expressing this commonality is to say that philosophers as diverse as William James and Friedrich Nietzsche, Donald Davidson and Jacques Derrida, Hilary Putnam and Bruno Latour, John Dewey and Michel Foucault, are antidualists... they are trying to shake off the influences of the peculiarly metaphysical dualisms which the Western philosophical tradition inherited from the Greeks: those between essence and accident, substance and property, and appearance and reality. They are trying to replace the world pictures constructed with the aid of these Greek oppositions with a picture of a flux of continually changing relations (Rorty 1999, 47).

Rorty describes this sort of view as "anti-metaphysical" because, following Heidegger, he regards all metaphysics as concerned with essences: "all Platonism is metaphysics and all metaphysics is Platonism" (ibid., 48). But notice how strongly Rorty's view echoes James's radical empiricism and Peirce's triadism. Rorty urges that we reject the metaphysical distinction between "intrinsic" and "extrinsic" properties or relations. Contra Rorty, I would argue that this amounts to a *metaphysical* anti-essentialism.

In turn, Rorty argues that this anti-essentialist move has consequences for how we think about language, thought, and perception. On his view, language and thought do not represent objects (or their essences), but are simply ways of getting into relations with those objects. Sentences about those objects are not to be distinguished as "true" or "false" but rather as more or less useful tools for acting with those objects.

3.3.5   Metaphysical Lessons for and from Perspectivism

Returning to the point discussed in the introduction: perspectivism starts from what the pragmatist might consider a problematic metaphor—that of vision, passively looking at something from a certain viewpoint. From this way of thinking about it, perspectives are inevitably partial, limited views of *the real thing*. The very underlying metaphor presupposes an aperspectival underlying reality, in which the real object exists in view of the various perspectives at hand. Not surprisingly, then, the metaphysical realist reacts to the perspectivist by pointing out that perspectivally partial knowledge is either merely partial knowledge, or no knowledge at all.

The pragmatist gives us an alternative to the metaphysical realist background which does not amount to mere quietism. And indeed, the pragmatist also offers us a way of thinking about perception and cognition that evades the spectator theory of knowledge and the centrality of the appearance/reality distinction. In "The Reflex Arc Concept in Psychology" (1896, EW 5:96ff), Dewey provides a powerful alternative to stimulus-response psychology, a precursor to modern-day ecological and enactive theories of perception (Gallagher 2014). According to this view, perception is not a matter of passive spectating, but actively engaging. It does not create a partial copy (an "appearance") of the real object in the mind, but is part of an ongoing circuit of sensorimotor engagement in the world. At every moment, the object, environment, and agent are interacting and reconstructing each other. Such a theory of perception could suitably recover the metaphor of "perspective" at the heart of perspectivist epistemology.

3.4   Pragmatist Standpoint Theories and Cultural Pluralism

Unfortunately neglected in much of the perspectivist literature is the range of human, social, and cultural difference we find in the world. Many perspectivists have, implicitly or explicitly, limited the range of perspectives to a "safe" subset, such as alternative scientific measurement techniques or scientific paradigms (e.g., Giere 2006). What's more, the examples are drawn from mainstream, modern, *Western* (or *Northern*) science. They ignore whether existing or possible, culturally distinctive, knowledge-making projects and practices might really count as *knowledge* or even as *science* (Harding 1994, 1998).

Perspectivists have yet to deeply consider how to incorporate a wider range of perspectives into science or what we are willing to regard as knowledge, such as women's perspectives, non-white perspectives, working class or economically disadvantaged perspectives, non-Western or non-Eurocentric perspectives, traditional religious perspectives, postcolonial perspectives, indigenous perspectives, and so on. Some of the American pragmatists, because they were reckoning with problems of public knowledge and social action in the context of America's highly pluralistic democracy, wrestled directly with how to approach such perspectives, think about their role in our knowledge system, and think about how to cooperate and integrate perspectives where needed.

3.4.1 Addams' Pluralistic Standpoint Epistemology

Departing from the more metaphysical ideas of the classical pragmatists, we can return to epistemic concerns, focusing on Jane Addams' version of standpoint epistemology, which also takes us into social and political concerns. Addams was a founding figure in the social settlement movement, where progressive social activists and reformers moved into "settlement houses" in poor, immigrant, or otherwise oppressed urban communities and provided services to the neighborhood. Addams co-founded Hull House in Chicago, and her experiences here formed a basis for her work as a public philosopher, a sociologist, a social worker, and a peace activist, for which she won the Nobel Peace Prize. Addams was born into a life of relative wealth and privilege, and her experiences in the neighborhood of Hull House formed an important part of her thinking about standpoint epistemology.

Addams based much of her philosophical and popular thought on her firsthand experience through Hull House. She was also sensitive to the ways that her understanding of those experiences might differ from the other people in the neighborhood whose social position was so different from her own. She adopted a strategy to address this problem:

> I never addressed a Chicago audience on the subject of the Settlement and its vicinity without inviting a neighbor to go with me, that I might curb my hasty generalization by the consciousness that I had an auditor who knew the conditions more intimately than I could hope to do (Addams 1910, 96).

As Maurice Hamington says in reference to this practice, "Addams did not try to arrive at universal moral truths but recognized that the standpoint of Hull House neighbors mattered" (Hamington 2018). More generally, in her writings on problems of labor, charity, poverty, and oppression, she constantly gave voice to the concerns and experiences of the oppressed people she encountered, recognizing that the standpoint of the oppressed was often more revealing about social conditions than the theories of privileged academics: "no one so poignantly realizes the failures in the social structure as the man at the bottom, who has been most directly in contact with those failures and has suffered most" (Addams 1910, 183).

Addams also held that it was crucial to bridge different standpoints through *sympathetic understanding* and *reciprocal responsibility* for the well-being of others. Addams points to the flaw in the charity-worker or the employer, "...when he is good 'to' people rather than 'with' them, when he allows himself to decide what is best for them instead of consulting them" (Addams 1902, 70). We must instead approach other people in attempt at social understanding and take responsibility for their welfare. These are crucial, in Addams' conception, to social science, social work, and social ethics.

3.4.2 *Du Bois on Race Consciousness*

W.E.B. Du Bois, as a historian and an American pragmatist philosopher, focused primarily on issues of race. He developed a conception, *double-consciousness*, which in some respects relates to standpoint theory, but uniquely addresses the condition of African-Americans. In *The Souls of Black Folk*, Du Bois writes:

> After the Egyptian and Indian, the Greek and Roman, the Teuton and Mongolian, the Negro is a sort of seventh son, born with a veil, and gifted with second-sight in this American world,—a world which yields him no true self-consciousness, but only lets him see himself through the revelation of the other world. It is a peculiar sensation, this double-consciousness, this sense of always looking at one's self through the eyes of others, of measuring one's soul by the tape of a world that looks on in amused contempt and pity. One ever feels his two-ness,—an American, a Negro; two souls, two thoughts, two unreconciled strivings; two warring ideals in one dark body, whose dogged strength alone keeps it from being torn asunder (Du Bois 1903, 3).

Standpoint theorists have tended to emphasize the epistemic advantage of the oppressed. Women understand both their own experience and they imbibe the official dogmas of the patriarchy; the proletariat knows both the ideology of the capitalist class and their own experiences of alienation and exploitation; etc. In *Souls* Du Bois emphasizes the way that this double-consciousness is itself alienating: oppressed folk know not only what white supremacists think about them, but they learn to measure themselves by those standards.

James Baldwin described double-consciousness in a particularly vivid way:

> In the case of the American Negro, from the moment you are born every stick and stone, every face, is white. Since you have not yet seen a mirror, you suppose you are, too. It comes as a great shock around the age of 5, 6, or 7 to discover that the flag to which you have pledged allegiance, along with everybody else, has not pledged allegiance to you. It comes as a great shock to see Gary Cooper killing off the Indians, and although you are rooting for Gary Cooper, that the Indians are you.

> It comes as a great shock to discover that the country which is your birthplace and to which your life and identity has not, in its whole system of reality, evolved any place for you ("The American Dream and the American Negro" 1965, Baldwin 1998, 98:714–15).

In a white supremacist culture, every facet of the media reflects white supremacist ideology. The protagonist in nearly every story is white. And even though today we have made improvements to the representation of other identities and perspectives, still the majority of literature, of history, of philosophy reflects white people and white perspectives.

The passage above from Du Bois's *Souls of Black Folk* also emphasizes the "gift of second-sight" that double-consciousness entails. Du Bois emphasizes this element more significantly in later writings, particularly in "The Souls of White Folk" in *Darkwater* (1920):

> Of [The Souls of White Folk] I am singularly clairvoyant. I see in and through them. I view them from unusual points of vantage. Not as a foreigner do I come, for I am native, not foreign, bone of their thought and flesh of their language. Mine is not the knowledge of the traveler or the colonial composite of dear memories, words and wonder. Nor yet is my knowledge that which servants have of masters, or mass of class, or capitalist of artisan. Rather I see these souls undressed and from the back and side. I see the working of their entrails. I know their thoughts and they know that I know. This knowledge makes them now embarrassed, now furious! (Du Bois 1920, 29).

Du Bois claims to know white folks, in a sense, more intimately than they know themselves. In part this is because, growing up in America, he is fed white supremacist ideology from early childhood, while at the same time, he sees the arbitrary lies at bottom of this ideology. Du Bois argued that writers like himself, in order to address racial problems, had to lead (white?) readers "within the Veil" so they could have a better understanding of the black experience (Gooding-Williams 2018). For Du Bois, too, then, something like *sympathetic understanding* was necessary for addressing racial problems. Standpoint theory and double-consciousness differ from some perspectivist views in that, while it treats knowledge as perspectival, it does not see all social perspectives as created equal.

3.4.3   Cultural Pluralism in Kallen, Locke, and Follett

Horace Kallen is perhaps better known for his essay, "Democracy versus the Melting Pot", in which he criticized the "melting pot" metaphor for the incorporation of immigrants into American society as, effectively, being a form of destructive cultural assimilation disguised by a thin veneer of tolerance. The idea was that immigrants should assimilate to a "common" American culture than in practice was simply the culture of the Anglophone majority. By contrast, Kallen argued that "*cultural pluralism*" was the foundation for the growth of American culture. "In manyness, variety, differentiation, lies the vitality of such oneness as they may compose" (Kallen 1924, 35). In a similar vein, Mary Parker Follett writes, "[t]he hope of democracy is in its inequalities" (Follett 1918, 139), by which she means, in the differences between the citizens.

For both Kallen and Follett, cultural pluralism had to be *integrative*, or melioristic in the sense of James and Addams, rather than a view where separate cultures keep to themselves. As Follett put it, "[u]nity, not uniformity, must be our aim. We attain unity only through variety. Differences must be integrated, not annihilated, nor absorbed" (Follett 1918, 39).[2] Kallen asked us to replace the metaphor of the melting pot with the metaphor of the orchestra:

> As in an orchestra, every type of instrument has its specific timbre and tonality, founded in its substance and form; as every type has its appropriate theme and melody in the whole symphony, so in society each ethnic group is the natural instrument, its spirit and culture are its theme and melody, and the harmony and dissonances and discords of them all make the symphony of civilization (Kallen 1924, 116–17).

Kallen qualified this metaphor by insisting that, unlike a symphony that is written ahead of time to be played by the orchestra, "in the symphony of civilization the playing is the writing" (ibid.). It is less an orchestra than a jazz ensemble, then.

Kallen acknowledged that he developed his conception of cultural pluralism in conversations with his African-American student, Alain Locke, who would go on to be called "Father of the Harlem Renaissance" (see McKenna and Pratt 2015; Carter 2012).[3] Locke was particularly interested in connecting cultural pluralism and value theory. He sometimes referred to his value theory as "value relativism", though his view did not contain the ultimate incommensurability often associated with the term "relativism". For Locke, cultural pluralism and value relativism largely involved the acknowledgement, on the one hand, that values are culturally dependent, and on the other hand, that cultural exchange should be approached from "*The principle of cultural reciprocity*" rather than a hierarchical relation between cultures (Harris 1991, 73).

The relativism was mitigated by the functional equivalences between values, "limited cultural convertibility" (ibid.), and the pervasiveness of the transvaluation of values, i.e., that all valuation evolves reevaluating and transforming our values (Harris 1991, 31ff.). While different cultures might value different things as "sacred", all cultures value *some* things as sacred. While what our religious values are may be culturally relative, all (or most) cultures have religious values. These kind of functional equivalences enable "limited cultural convertibility", i.e., communication across cultures about values independent of their specific cultural contents. Locke thought transvaluation, the reevaluation and transformation of our values, was common both for individuals and society. All of these benefitted from cultural exchanges under conditions of reciprocity.

---

[2] Follett took pains to make sure that her account of "integration" was not misunderstood along assimilationist lines. In a footnote to this passage, she wrote:

I therefore give here a list of words which can be used to describe the genuine social process and a list which gives exactly the wrong idea of it. Good words: integrate, interpenetrate, interpermeate, compenetrate, compound, harmonize, correlate, coordinate, interweave, reciprocally relate or adapt or adjust, etc. Bad words: fuse, melt, amalgamate, assimilate, weld, dissolve, absorb, reconcile (if used in Hegelian sense), etc. (Follett 1918, 39).

[3] Although compare Kallen's rather condescending memorial address, "Alain Locke and Cultural Pluralism" (1957).

3.4.4   Cultural Lessons for and from Perspectivism

Perspectivism in epistemology and philosophy of science recognizes the partiality and the plurality of claims to knowledge. But it does not always acknowledge the range of perspectives, particularly perspectives deriving from other cultures, including non-Western and indigenous knowledge traditions. Here perspectivism could learn from the postcolonial and multicultural approaches to science as explored, for example, by Sandra Harding (Harding 1994, 1998). Cultural pluralism reminds perspectivism and pragmatism that they have rich resources for thinking beyond a narrow-minded Eurocentrism to a position where those who have been pushed to the epistemic margins are given a voice and have their contributions to knowledge respected.

Another element highlighted by these standpoint theory and cultural pluralist approaches from the broader American pragmatist tradition is the relation of power and oppression to knowledge. Pragmatists and American philosophers like Addams, Du Bois, Kallen, and Locke join with other feminists, Marxists, and critical race theorists in developing standpoint epistemology and theories of double-consciousness that remind us that not only are not all perspectives created equal, but the perspective of oppressed groups has a special relevance to many questions of knowledge. It also reminds us of the ethical requirements of approaching different perspectives in remaining humble, practicing reciprocity, and being aware of the dimensions of our own perspectives' social privileges.

Finally, this discussion points towards the need to build bridges and reciprocity between perspectives. We should not be satisfied with different perspectives in isolation from one another. That does not mean assimilation of one perspective to another. Rather, as Follett would insist, it points towards a non-dominationist account of *integration* of perspectives. To continue the visual metaphor, integrating perspectives into a kind of *stereoscopic* vision reveals dimensions of understanding not available to an isolated perspective. That advantage is destroyed by reduction of one perspective to another or both to a third. This suggests a fruitful connection between perspectivism and what Sandy Mitchell has called "integrative pluralism" (Mitchell 2002, 2003), acknowledging that integration is a difficult and piece-meal task.

3.5   Conclusion

This paper has provided merely an overview of potential points of connection between contemporary perspectivism and the American pragmatist tradition. It is my hope that contemporary perspectivists might find valuable resources within the pragmatist tradition for developing their views, and that this chapter might spark mutual insight and appreciation between contemporary perspectivists and pragmatists.

Bibliography

Addams, J. (1902). In C. H. Seigfried (Ed.), *Democracy and social ethics*. Urbana: University of Illinois Press (2002).

Addams, J. (1910). *Twenty years at Hull-House*. Urbana: University of Illinois Press.

Alexander, T. M. (2014). Linguistic pragmatism and cultural naturalism: Noncognitive experience, culture, and the human Eros. *European Journal of Pragmatism and American Philosophy, 6*(2).

Bain, A. (1859). *The emotions and the will*. New York: Longman's Green.

Baldwin, J. (1998). *Collected essays* (Vol. 98). New York: Library of America.

Capps, J. (2018). Did Dewey have a theory of truth? *Transactions of the Charles S. Peirce Society, 54*(1), 39.

Capps, J. (2019). The pragmatic theory of truth. In Edward N. Zalta (ed.), *The Stanford encyclopedia of philosophy*. Summer 2019. https://plato.stanford.edu/archives/sum2019/entries/truth-pragmatic/; Metaphysics Research Lab, Stanford University.

Carter, J. A. (2012). Alain Leroy Locke. In Edward N. Zalta (ed.), The Stanford encyclopedia of philosophy, Summer 2012. https://plato.stanford.edu/archives/sum2012/entries/alain-locke/; Metaphysics Research Lab, Stanford University.

Dewey, J. (1910). The postulate of immediate empiricism. In *The Influence of Darwin on Philosophy and Other Essays* (pp. 226–241). New York: Henry Holt and Company.

Dewey, J. (1917). The need for a recovery of philosophy. In *Creative intelligence: Essays in the pragmatic attitude* (pp. 3–69). New York: Holt.

Dewey, J. (1922). Syllabus: Types of philosophic thought. In J. A. Boydston (ed.), *The middle works of John Dewey, 1899–1924* (Vol. 13: 1921–1922 Essays). Carbondale: Southern Illinois University Press.

Dewey, J. (1933). *How we think: A restatement of the relation of reflective thinking to the educative process*. Edited by Jo Ann Boydston. Vol. 8. The Later Works of John Dewey. Southern Illinois UP, 1986/2008.

Dewey, J. (1938). *Logic: The Theory of Inquiry*. Edited by Jo Ann Boydston. Vol. 12. The Later Works of John Dewey. Southern Illinois: UP, 1991.

Dewey, J. (1941). Propositions, warranted assertibility, and truth. *The Journal of Philosophy, 38*(7), 169–186.

Du Bois, W. E. B. (1903). *The souls of black folk: Essays and sketches*. Chicago: A. C. McClurg & Co. http://bartelby.com/114.

Du Bois, W. E. B. (1920). *Darkwater: Voices from Within the Veil*. New York: Harcourt, Brace; Howe.

Fisch, M. H. (1954). Alexander Bain and the genealogy of pragmatism. *Journal of the History of Ideas, 15*(3), 413–444.

Follett, M. P. (1918). *The new state: Group organization the solution of popular government*. New York: Longmans, Green.

Gallagher, S. (2014). Pragmatic interventions into enactive and extended conceptions of cognition. *Philosophical Issues, 24*(1), 110–126.

Giere, R. N. (2006). *Scientific perspectivism*. Chicago: University of Chicago Press.

Gooding-Williams, R. (2018). W.E.B. Du Bois. In Edward N. Zalta (ed.), *The Stanford Encyclopedia of Philosophy*, Summer 2018. https://plato.stanford.edu/archives/sum2018/entries/dubois/; Metaphysics Research Lab, Stanford University.

Haack, S. (1982). Descartes, Peirce and the cognitive community. *The Monist, 65*(2), 156–181.

Hamington, M. (2018). Jane Addams. In Edward N. Zalta (ed.), *The Stanford Encyclopedia of Philosophy*, Summer 2018. Metaphysics Research Lab, Stanford University. https://plato.stanford.edu/archives/sum2018/entries/addams-jane/.

Harding, S. G. (1994). Is science multicultural?: Challenges, resources, opportunities, uncertainties. *Configurations, 2*(2), 301–330.

Harding, S. G. (1998). *Is science multicultural?: Postcolonialisms, feminisms, and epistemologies*. Bloomington: Indiana University Press.

Harris, L. (Ed.). (1991). *The philosophy of Alain Locke*. Philadelphia: Temple University Press.

Houser, N., & Kloesel, C. J. W. (1992). *The essential Peirce: Selected philosophical writings* (Vol. 1). Bloomington: Indiana University Press.

James, W. (1890). *The principles of psychology*. New York: H. Holt.

James, W. (1892). *Psychology: Briefer course*. New York: Henry Hol.

James, W. (1896). The will to believe. *The New World, 5*, 327–347.

James, W. (1899). *Talks to teachers on psychology: And to students on some of Life's ideals*. New York: H. Holt.

James, W. (1907). *Pragmatism: A new name for some old ways of thinking*. New York: Longmans.

James, W., & Perry, R. B. (1912). *Essays in radical empiricism*. In Vol. 3. The works of William James. Cambridge, MA: Harvard University Press, 1976.

Kallen, H. M. (1924). *Culture and democracy in the United States*. New York: Boni; Liveright.

Kallen, H. M. (1957). Alain Locke and cultural pluralism. *The Journal of Philosophy, 54*(5), 119–127.

Magnus, P. D. (2013). What scientists know is not a function of what scientists know. *Philosophy of Science, 80*(5), 840–849.

McKenna, E., & Pratt, S. L. (2015). *American philosophy: From wounded knee to the present*. New York: Bloomsbury Press.

Misak, C. J. (2004). *Truth and the end of inquiry: A Peircean account of truth*. Oxford: Clarendon Press.

Mitchell, S. D. (2002). Integrative Pluralism. *Biology and Philosophy, 17*(1), 55–70.

Mitchell, S. D. (2003). *Biological complexity and integrative pluralism*. Cambridge: Cambridge University Press.

Peirce Edition Project, ed. (1998). *The essential Peirce: Selected philosophical writings* (Vol. 2). Bloomington: Indiana University Press.

Putnam, H. (1992). The permanence of William James. *Bulletin of the American Academy of Arts and Sciences, 46*(3), 17–31.

Rorty, R. (1999). *Philosophy and social Hope*. London: Penguin Books.

Chapter 4
Hilary Putnam on Perspectivism and Naturalism

Mario De Caro

Abstract In this chapter, I analyze the different views that Hilary Putnam developed during the seven decades of his brilliant philosophical career in order to address the question of realism (metaphysical realism, internal realism, quietism, and liberal naturalism). I also argue that the view Putnam defended at the end of his career was the most solid and consistent, and that such view could offer a useful inspiration to the advocates of perspectivism.

Keywords Hilary Putnam · Realism · Liberal naturalism · Perspectivism

4.1 Preamble

When Michela Massimi invited me to participate in the Edinburgh conference on perspectivism from which this volume derives, she asked me to give a talk about Hilary Putnam's "internal realism" (which was the conception that he defended between 1976 and 1990). Michela's supposition, which she wanted to test, was that "internal realism" can be seen as an early version of perspectivism, and a very interesting one. I agreed with her supposition, but I added that perspectivists should be more interested in "liberal naturalism", the view adopted by Putnam in his last years, since the latter view—granted it preserves the important perspectivist insights of internal realism—was preferable to internal realism for a number of independent reasons. In this chapter, I will develop and defend this interpretation of Putnam's work.

M. De Caro (✉)
Università Roma Tre, Roma, Italy

Tufts University, Medford (MA), USA
e-mail: mario.decaro@uniroma3.it

A. Crețu, M. Massimi (eds.), *Knowledge from a Human Point of View*,
Synthese Library 416, https://doi.org/10.1007/978-3-030-27041-4_4

4.2   For and against Metaphysical Realism

At the very beginning of his career, Putnam defended a form of physicalist monism, by articulating the logical-positivist thesis of the 'Unity of Science' (Oppenheim & Putnam 1958). He did so by claiming the (in principle) reducibility of the concepts and laws of higher-level sciences to the concepts and laws of the lower-level sciences—where microphysics represented the most fundamental level to which the others were in principle reducible. Much later, Putnam (2016b, 126) wrote that according to that view,

> Every phenomenon that can be explained by 'higher-level' sciences such as psychology and sociology could in principle be explained by 'lower-level' sciences, and ultimately by physics.

Putnam and Oppenheim did not claim that that thesis was justified *a priori*. Rather, they claimed that it was inspired by "a pervasive trend within scientific enquiry … notwithstanding the simultaneous existence (and, of course, legitimacy) of other, even incompatible trends" (Oppenheim and Putnam 1958, 4). Thus, already in his early physicalist years Putnam put the analysis of the concrete scientific practice at the centre of his philosophical project—an early sign of his dedication to the spirit and themes of the pragmatist tradition.

Shortly after that work, and until his death in 2016, Putnam changed deeply his philosophical attitude. More specifically, he began to criticize strongly *metaphysical realism*, a conception of which his early 'Unity of Science' article was a particular version.[1] According to metaphysical realism, only one true and complete description of the world exists, which is typically regarded as being offered by the natural sciences, especially physics (for a defence of this view, in different fashions, please see Smart 1978; Field 1992; Papineau 1996; Kim 2005; Stoljar 2010). A contemporary version of the view is offered by Alex Rosenberg (2013, p. 19):

> What is the world really like? It's fermions and bosons, and everything that can be made up of them, and nothing that can't be made up of them. All the facts about fermions and bosons determine or 'fix' all the other facts about reality and what exists in this universe or any other if, as physics may end up showing, there are other ones. In effect, scientism's metaphysics is, to more than a first approximation, given by what physics tell us about the universe. The reason we trust physics to be scientism's metaphysics is its track record of fantastically powerful explanation, prediction and technological application. If what physics says about reality doesn't go, that track record would be a totally inexplicable mystery or coincidence.[2]

For the rest of his career, Putnam saw metaphysical realism as a dogmatic and philosophically pernicious conception and developed several arguments against it.

---

[1] In a lecture delivered at Oxford in 1960, and published as (1975a), Putnam offered another defence of the basic principles of metaphysical realism.

[2] It is interesting that, in order to describe his view, Rosenberg employs the term 'scientism', which is normally used with derogatory connotations.

The first attack was in the 1960s, when Putnam developed 'Computational Functionalism,' an extremely influential conception of the mind-body problem, which was pluralist in ontology, since he accepted the existence of irreducible mentalistic properties. According to that view, mental functions are 'hardwired' in the brains of the speakers so that the relation between the mind and the brain is analogous to the relation between the software and the hardware. In this framework, mental properties depend for their existence on physical properties, but could not be reduced to them; consequently, one cannot investigate mental properties only by appealing to the conceptual tools offered by physics or by any of the natural sciences. Expanding on this doctrine, during this period Putnam explicitly rejected the Unity of Science and developed (at the same time as Jerry Fodor in 1965, but independently from him) a pluralistic view that Ned Block (1997, 108) would later call the 'Many Levels Doctrine'.[3] From this perspective, "nature has joints at many different levels, so at each level there can be genuine sciences with their own conceptual apparatus, laws and explanations" (Block 1997, 108). An example that clarified that view was offered by Putnam in his influential article "Reductionism and the Nature of Psychology" (Putnam 1973). Why does a solid rigid round peg, which is a little less than 1 inch in diameter, fit through a round square of 1 inch in diameter but does not fit in a square hole with a diagonal of 1 inch? The correct answer to this question, according to Putnam, cannot be given by appealing to the physical (lower-level) properties of the peg and holes, but only to the geometrical (high-level) properties; this shows that different levels of reality are composed by different and mutually irreducible properties. At this stage of his career, therefore, Putnam began to believe in the irreducibility of the mental to the physical—a cornerstone of the common-sense view of the world—and, more generally, he saw several reasons for thinking that physics does not delimit the boundary of reality and knowledge, and that conceptual, epistemological and ontological pluralism is in order.

Throughout most of his career Putnam thought that the language of physics cannot account for everything existing since some real features of the world cannot be described, even less explained, with the conceptual tools of that science. As Putnam (65) wrote:

> The world cannot be completely described in the language game of theoretical physics, not because there are regions in which physics is false, but because, to use Aristotelian language, the world has many levels of forms, and there is no realistic possibility of reducing them all to the level of fundamental physics.

What are the "levels of forms" that are not reducible to the level of fundamental physics, then? One of Putnam's favourite examples was literary criticism: how could the natural sciences render the sense of the kind of understanding that is offered to us when a critic interprets, say, a play by Shakespeare: "what exactly does it mean to 'apply' the supposed 'methods of the natural sciences' to Julius Caesar?"

---

[3] See the essays collected in Putnam (1975b).

(Putnam 2015, 312). Putnam offered a number of other examples from spirituality,[4] mathematics,[5] and ethics.[6]

In his late years, Putnam was a strong advocate of pluralism at the conceptual, the epistemological, and the ontological level. In this light, he developed various versions of realism that simultaneously accepted the approximate and revisable truth of the different views of the world offered by the natural, social and human sciences, common sense, and the arts. More specifically, Putnam articulated his pluralism in two ways. First, he described a phenomenon that he called 'conceptual relativity', for which some theories can be cognitively equivalent, even if *prima facie* they appear incompatible. (It would have been better if this phenomenon had been called 'cognitive equivalence', since Putnam's original term may suggest a connection with relativism and antirealism that is entirely inappropriate). An example of conceptual relativity frequently offered by Putnam concerns mereological sums: a world of three individuals could also be seen as composed by seven objects: the three individuals and all their possible sums. However, Putnam also offers less convoluted examples of conceptual relativity. In fact, he wrote, in some scientific fields, such as mathematical physics, this phenomenon is ubiquitous:

> To take an example from a paper with the title 'Bosonization as Duality' that appeared in *Nuclear Physics* B some years ago, there are quantum mechanical schemes some of whose representations depict the particles in a system as bosons while others depict them as fermions. As their use of the term 'representations' indicates, real live physicists—not philosophers with any particular philosophical axe to grind—do not regard this as a case of ignorance. In their view, the 'bosons' and 'fermions' are simple artifacts of the representation used. But the system is mind-independently real, for all that, and each of its states is a mind independently real condition that can be represented in each of these different ways. And that is exactly the conclusion I advocate…. [These] descriptions are both answerable to the very same aspect of reality… they are 'equivalent descriptions'. (Putnam 2012a, 63–64).

The second pluralistic issue that Putnam (2005/2012, 64–65) stressed concerned another (more meaningful) phenomenon, that of 'conceptual pluralism', i.e., the fact that for understanding the different levels of reality we need a plurality of mutually irreducible but not incompatible conceptual systems. A favorite example by Putnam was that, depending on our interests, we can correctly and usefully describe a chair in the alternative languages of carpentry, furniture design, geometry, and etiquette. Let's give another example. A person had a heart attack, and one has to explain what happened: this can be done in different ways. One can offer a physiological explanation or refer to the dietary habits of the person or to the fact that she

---

[4] Unsurprisingly, considering his readiness to change intellectual positions, Putnam's views about religion went through different phases. After an atheist period, starting in the 1980's he went through a theistic phase, to conclude with a more nuanced view, in which spirituality was disconnected from the supernatural (see Putnam 2008).

[5] See Putnam (2012, part II).

[6] Putnam (2004). Putnam could not disagree more with attempts to naturalize mathematics, such as Maddy (1997), or to show that its statements are false since mathematics cannot be naturalized, such as Field (1980, 1989).

did not exercise enough or that ate bad food; moreover, a geneticist could talk about the person's family predispositions, a chemist of the reactions that produced the heart attack, and so on. Each of those explanations is legitimate and can be useful in its specific way (depending on which kind of explanation we are looking for), without being reducible to the others nor incompatible with them; so to speak, there is no fundamental and unifying theory of what being a chair is or why a heart attack can happen. And this is true of a vast amount of entities and situations (possibly all of them, with the exception of the entities of microphysics), since they can be described in different ways not just because of conceptual relativity, but also because things have different properties that belong to different ontological regions.

This view is at odds with the Quinean idea, which has inspired legions of naturalistically-inclined philosophers, that only our 'first-grade conceptual system' (that is, the view offered by the natural sciences) can offer a truthful description of the world. Putnam wrote:

> The heart of my own conceptual pluralism is the insistence that the various sorts of statements that are regarded as less than fully rational discourse, as somehow of merely "heuristic" significance, by one or another of the "naturalists" (whether these statements be ethical statements or statements about meaning and reference, or counterfactuals and statements about causality, or mathematical statements, or what ever) are bona fide statements, "as fully governed by norms of truth and validity as any other statements", as James Conant has put it (Putnam 2012, 112).

4.3   Internal Realism

With the exception of his early metaphysical-realist years, during his entire philosophical career Putnam tried to develop a satisfying meta-philosophical conception that could give unity to the aforementioned pluralistic insights. The first complete attempt, on which he worked between 1976 and 1990, was the so-called "internal realism". During this period, under the inspiration of Immanuel Kant, C.S. Peirce, and Michael Dummett, Putnam thought, and argued forcefully, that the only satisfying way of responding to metaphysical realism without appealing to supernatural entities and explanations was to adopt an epistemic view of truth—that is, a view in which truth is intended as warranted assertibility in idealized epistemic conditions (Putnam 1981).

One of the main rationales for internal realism was an argument Putnam first offered in 1977, when he delivered a talk to the Association for Symbolic Logic entitled "Models and Reality" (Putnam 1980). In that talk, he presented the so-called "model-theoretic argument" against metaphysical realism. By appealing to the Löwenheim–Skolem theorem, the argument aimed at showing that a metaphysical realist cannot fix the intended interpretation of his theory of the world (the only way to do it, he argued, would be by using supernatural powers). This is how, in an article written shortly before his death, he reconstructed his view of that period:

> According to the position I defended in "Models and Reality", … [the idea of] truth (about things outside my brain/mind) as *correspondence to the way things are* is empty (empty

because, to use [Charles] Travis's words, there is no unique set of facts as to what, without the boundary, instances what). But we could save the notion of truth, I claimed: truth is (idealized) "rational acceptability". But rational acceptability was supposed to mean acceptability relative to all the facts available to the subject's brain/mind about the subject's own *sense data* (Putnam forthcoming).

From this perspective, the correspondence theory of truth was doomed. Thus, he adopted the view according to which truth is verifiability in ideal epistemic conditions. But how "ideal" should the ideal epistemic conditions be? On this topic, Putnam changed his mind and that was one of the reasons why eventually, in 1990, he abandoned internal realism. In Putnam (1982, 55) he had given a very idealized interpretation to that notion:

'Epistemically ideal conditions' ... are like 'frictionless planes': we cannot really attain epistemically ideal conditions, or even be absolutely certain that we have come sufficiently close to them. But frictionless planes cannot really be attained either, and yet talk of frictionless planes has 'cash value' because we can approximate them to a very high degree of approximation.

However, in Putnam (1990, vii), his attitude had deeply changed and the interpretation of the "ideal" conditions was much closer to common sense:

If I say 'There is a chair in my study", an ideal epistemic situation would be to be in my study with the lights on or with daylight streaming through the window, with nothing wrong with my eye-sight, with an unfocused mind, without having taken drugs or been subjected to hypnosis, and so forth, and to look and see if there is a chair there.

The time was ripe for abandoning internal realism. Adopting such a commonsensical interpretation of the ideal epistemic conditions opened in fact the way to some striking counterexamples to the idea that truth coincides with verifiability in ideal epistemic conditions. Putnam's favorite example of why truth is not epistemically constrained is the conjecture "There is no life outside the earth" – which may well be true but, in case it is, would be unverifiable even in ideal epistemic conditions. In abandoning the epistemic view of truth, however, Putnam realized that he did not need to go antirealist in order to refuse the dogmatic view that he had called "metaphysical realism". His new aim was in fact to develop "a modest non-metaphysical realism squarely in touch with the results of science" (Putnam 2004, 286, n. 1).

4.4 Beyond Internal Realism

Between the end of the 1980's and the 1990's Putnam had additional reasons for abandoning internal realism. One such reason was the so called "no-miracles argument", which he had developed in 1973, but whose relevance he fully appreciated only later (Putnam 1975a, 2012b). This argument is based on the idea that the only way of explicating the great explanatory and predictive success of the best theories of modern science is to acknowledge that these theories are true (or approximately true) in regard to the natural world and refer to real entities, even when those are unobservable. From the point of view of antirealism, on the contrary, the fact that science works so well in offering comprehensive explanations and extremely precise predictions of observable phenomena does seem a sheer miracle. Consequently, according to Putnam, we should take our best scientific theories as true or approximately true and the entities those theories refer to as real – even though, of course, that does not mean that our theories cannot be false (Putnam always defended fallibilism with great conviction).

At first, Putnam insisted that the no-miracles argument was compatible with internal realism. The reason was that in his earlier view that was seen by Putnam as supporting entity-realism, not theory-realism. So one could accept the existence of atoms without endorsing the idea that our atomistic theories (which cannot be directly verified) are true: they should be taken merely as useful heuristic tools, as the instrumentalist have always done. In the 1990's, however, Putnam started to see this point of view as unsatisfying: why should we accept as granted the existence of atoms which we derive from our theories, while thinking that those same theories cannot be true? Indeed the reason for that, he thought, was simply the mistaken assumption that only internal reason could offer a feasible alternative to metaphysical realism. However, scientific realism does *not* imply metaphysical realism: one can believe in the truth of our best physical theories (and in the existence of the entities whose existence they presuppose) without endorsing the other, much stronger thesis that physics can in principle account for everything existing. In that light, scientific realism can be (and, according to Putnam after 1990, should be) reconciled with conceptual, epistemic, and ontological pluralism.

Another reason why Putnam abandoned internal realism was that he did not think anymore that the aforementioned argument he had offered in "Models and Reality" was correct. As said, the argument tried to reconcile two views: the epistemic conception of truth and the thesis—defended via the Löwenheim–Skolem theorem—that there is no unique set of facts to which our thoughts can be said to refer. And in this, as Putnam (forthcoming) noticed later, internal realism collapsed on solipsism, a very suspicious philosophical view:

> … rational acceptability was supposed to mean acceptability relative to all the facts available to the subject's brain/mind about the subject's own *sense data*. But this is solipsism! [And in] *Reason, Truth and History* I tried to avoid solipsism with the aid of a counterfactual: truth is *what would be rationally acceptable if epistemic conditions were ideal!* But in "Models and Reality" I pulled the rug out from under myself (without noticing that I did), when I pointed out that counterfactuals are no help here. Thus appeal to counterfactuals cannot rule out any models at all unless the interpretation of the counterfactual idiom itself is *already* fixed by something beyond operational and theoretical constraints.

In another article written in his later years, Putnam (2012a, 80), summarized this criticism to internal realism:

> *Moral*: My "internal realism", far from being an intelligible alternative to a supposedly unintelligible Metaphysical Realism, can itself possess no *public* intelligibility. And the situation may be worse: Dummett is right to worry whether a verificationist account of understanding does not commit one to antirealism about the past. If it does, then, … even methodological solipsism collapses into hopeless paradoxes. The best way to show that the

realist position isn't just one of two equally tenable positions is to show that the verifica-
tionist account entails solipsism (and probably even entails a self-refuting antirealism about
the past). If this is right, then it clearly becomes vital to give an account of our capacity to
understand and use language that 'fits' with realism.

Seeing that internal realism implied solipsism gave Putnam another excellent reason
to give it up. This new reason was that in the early 1970s he had developed an externalist semantics, whose fatal consequences for internal realism he realized only
later.[7] Putnam appealed to the famous Twin Earth thought experiment in order to
justify the idea that thoughts are supposed to refer to unobservable entities (an idea
that involves the revolutionary thesis that, *contra* Frege, intensions do not determine
references—see Putnam 2016c, 208–209). In order for our words to have meaning
we *have* to refer to the external world—and even to the parts of the external world
that we are not able to access: solipsism was therefore proven wrong, and internal
realism with it.

Finally, also Putnam's views about the mind-body problem took Putnam toward
a more realistic direction, when he abandoned 'computational functionalism'
(according to which mental functions are hardwired in the brain of a speaker), in
favor of a view that he called 'long-armed functionalism'. This is a view of the mind
as a system of object-involving abilities that involve, from the start, the natural and
social environment in which a speaker is located. Several things have to be noted in
this respect. First, the fundamental reason as to why Putnam abandoned computational functionalism was that he realized it was incompatible with his semantic
externalism, according to which the relation between the thinkers and the environment they inhabit is necessary for constituting the content of at least some of their
thoughts. In this regard, Putnam wrote:

> I had to give up 'functionalism', ... that is, the doctrine that our mental states are just our
> *computational* states (as implicitly defined by a 'program' that our brains are hard-wired to
> 'run'), because that view is incompatible with the semantic externalism that years of think-
> ing about the topic of reference had eventually led me to develop. If, as I said in 'The
> Meaning of 'Meaning'', our intentional mental states aren't in our heads, but are rather to be
> thought of *as world-involving abilities*, abilities identified by the sorts of transactions with
> our environment that they facilitate, then they aren't identified simply by the 'software' of
> the brain (Putnam 2005/2012, 58).

Then, with regard to long-armed functionalism, he wrote:

> [it] is an antireductionist but naturalist successor to the original, reductionist, functionalist
> program. For a liberalized functionalist, there is no difficulty in conceiving of ourselves as
> organisms whose functions are, as Dewey might have put it, 'transactional', that is
> environment-involving, from the start (Putnam 2012).

Also in the philosophy of mind, then, Putnam (by abandoning the idea that the mind
is the sum of the computational states running on brain states) came to abandon the
idea of the primacy of the internal. In his new view the mind was, so to say, a set of

---

[7] The classic account of semantic externalism is offered in Putnam (1975c); see also Putnam
(2016c) for an account of the development of his views about externalism.

capacities that necessarily involve our transactions with the external world: also in this field, externalism had won its battle with internalism.

4.5   Liberal Naturalism

Internal realism had then to be abandoned. What kind of conception could incorporate Putnam's old anti-metaphysical views (conceptual pluralism, conceptual relativity, epistemic pluralism and ontological pluralism) with his new realist claims?

Putnam made two attempts to unify these views. First, in the 1990s he developed a form of Wittgensteinian quietism—that is, a skeptical attitude toward metaphysical problems (which should be 'dissolved' more than 'solved', since they are based on conceptual confusions). In this period, Putnam was influenced by Austin's direct realism and by McDowell's insistence on the conceptual independence of the 'realm of reason' from the 'realm of law'. However, as usual, also some pragmatists and perspectivist themes worked in the background of this phase of Putnam's philosophical development. Let us consider some of them.

The first pragmatist/perspectivist theme that inspired Putnam during that phase (and also later) was the idea that causality is an essentially intentional notion, since it is inextricably connected with our explanatory practices. In this respect, after approving John Haldane's saying that "there are as many kinds of causes as there are senses of 'because'",[8] Putnam wrote that "[c]ausality depends on the interests at stake when one asks the question: 'What is the cause of that?'".[9] It should be noted that with the term 'senses of 'because', Putnam means, in the spirit of pragmatism and perspectivism, "our ever expanding repertoire of explanatory practices" (Putnam 1999, 150; also Putnam & Putnam 2017). In this light, the combination of pluralism about explanation, on the one hand, and the conceptual link between explanation and causation, on the other hand, generates pluralism about causation.

In line with this thesis, Putnam claimed that the so-called 'principle of the physical causal closure of the world' should be rejected, as long as one takes it in one of its classic formulations: "If x is a physical event and y is a cause or effect of x, then y, too, must be a physical event" (Putnam 1999, 215). Given Putnam's pluralist and non-reductionist ontological attitude, physical events can indeed be caused by non-physical events that are irreducible to physical events—which is a form of downward causation. (It should be noted, however, that this does not mean that at the same time events cannot have physical causes as well). The crucial point, for Putnam, is that different causal explanations generalize to different classes of cases: and whether we are interested in an event as a member of one or another class is a completely context-relative question. For example, we can be interested in the physiological chain of events that ended in the movement of my hand; but we can

---

[8] Putnam (1999, 201, n. 17) writes that neither him nor Haldane could remember where the latter wrote the quoted phrase.

[9] Quoted in Putnam (1999, 77); also ibid. 137, and 149–150.

also be interested in the reasons for which I intentionally moved it. Neither of these causal chains have priority on the other since their respective interests are context-relative.

In this period, Putnam also explored in depth his old idea that between factual and evaluative statements there is no conceptual *dichotomy*, since between them there is only a (sometimes useful) *distinction*.[10] This is because, according to Putnam, values and normativity are ubiquitous: even scientists appeal to values—which can be epistemic or even aesthetic—in order to choose between cognitively equivalent theories. Going back to Galileo, for example, it should be remembered that his main reason for accepting the Copernican system was an aesthetic one.[11]

The last of Putnam's later views that we can mention here is *liberal naturalism*, which he saw as the general framework of most of the ideas he had held in his last years.

> I very much like the term 'liberal naturalism' which I first encountered in an important collection of essays edited by Mario De Caro and David Macarthur [2012] titled *Naturalism in Question*.... In their introduction to *Naturalism in Question*, De Caro and Macarthur emphasized that the liberal naturalism they advocate doesn't regard normative utterances as somehow 'second grade' or merely 'expressive', but neither does it countenance a Platonic realm of normative facts independent of human practices and needs. At the same time, it does not countenance Moorean quasi-mystical faculties of moral intuition. All this I like very much (Putnam 2015, 312–313).

Liberal naturalism is a metaphilosophy that advocates a pluralistic attitude both in ontology and epistemology on the basis of the ideas that not all the real features of the world can be reduced to the scientifically describable features and that the natural sciences are not the only genuine source of knowledge to which all the other apparent sources should hand over their epistemic pretensions. Still, a liberal naturalist cannot accept any entity in her ontology or any view in her epistemology that would contradict the current scientific worldview.[12]

Putnam's liberal naturalism seems to be a very promising perspective. As with any serious philosophical view, however, it faces several problems, which can be considered as parts of a big metaproblem, which can be called 'The Reconciliation Problem': what kind of relation is there between the accounts of the world offered by the natural sciences and those offered by the social sciences, common-sense, the arts, and spirituality? To be more specific: what is the relationship between the ontological realm studied by the natural sciences and the other ontological regions? Is that a relation of *supervenience* (and in this case, of which kind?), *emergence*,

---

[10] Putnam (1982, 2002, and 2011)

[11] "[The Ptolemaic system was] a monstrous chimera composed of mutually disproportionate members, incompatible as a whole. Thus however well the astronomer might be satisfied merely as a *calculator*, there was no satisfaction and peace for the astronomer as a *scientist*. And since he very well understood that although the appearances might be saved by means of assumptions essentially false in nature, it would be very much better if he could derive them from true suppositions" (Galilei 1632, 341; emphasis added).

[12] Putnam contributed to both De Caro and Macarthur (2004, 2010), which advocated liberal naturalism, and defended the latter view also in Putnam (2016a e 2016b).

grounding, incommensurability, or something else? Putnam was happy with global supervenience (according to which any world physically identical to ours would have the same non-physical properties of our world), but this is a controversial issue. And connected with this difficulty, another one immediately raises: what about the traditional problems of causal over-determination and the violation of the closure of the physical world?

These problems have to be addressed by the philosophers who want to legitimise the common-sense features of the world. In fact, they have the unavoidable task of showing how Sellars's manifest and scientific images of the world can co-exist when they are both taken as fully legitimate and non-hierarchically related (De Caro 2015). Despite these difficult problems, in his last years Putnam thought that liberal naturalism was the preferable view since, without appealing to any supernatural feature, it is much less revisionistic than the scientistic views mentioned at the beginning of this chapter regarding the features of the world that can be grasped by other forms of understanding, such as literature or common sense.

4.6 Perspectivism and Liberal Naturalism

Perspectivism is an attempt—explicitly inspired by Kant, Kuhnian relativism, and Putnam's internal realism—at finding a middle ground between scientific realism and antirealism. Michela Massimi (2016, 2017) offers excellent general introductions to this view, and so do many of the papers of this collection; thus, I will be brief on the details of this view and go directly to an evaluation.

Ronald Giere, one of the leading advocates of this view, claims that scientific observation, measurement, modelling, theorizing are all perspectival (that is, they depend on the human point of view) and therefore scientific knowledge as such is unavoidably contingent. Giere writes about truth as follows: "Truth claims are always relative to a perspective", since no theory "provides us with a complete and literally correct picture of the world itself" (2006, 81). He also adds:

> Full objectivist realism ("absolute objectivism") remains out of reach, even as an ideal. The inescapable, even if banal, fact is that scientific instruments and theories are human creations. We simply cannot transcend our human perspective (Giere 2006, 14–15).

It is unclear to me why, granting fallibilism (for which each empirical claim, individually taken, could be false), one could not accept the idea that in some fields our theories could reach the objective truth: in that case, of course, we could not be absolutely sure of the truth of those theories; however, they would still be objectively true. I therefore agree with Michela Massimi's view that Giere's perspectivalism fails to find a solid middle ground between realism and antirealism, since it is too unbalanced toward a form of antirealism. In particular, besides all the other objections against that family of views, Giere's view is also exposed to another criticism: is not there a risk that under Giere's view the objective external world becomes a new version of Kant's noumenal (and hence unknowable) world?

Massimi's version of perspectival realism is, in my view, a much more balanced view then Giere's, since, besides incorporating what one could get by, by buying into some forms of antirealism (that is, epistemic pluralism), it also does justice to the main realist intuitions. In my view, Massimi's perspectivism is (i) a legitimate form of realist naturalism; (ii) it's neutral (for what I know) between strict and liberal naturalism; and (iii) it's closer to the late Putnam's liberal naturalist realism than to his internal realism. One interesting passage in Massimi (2017, 170) on the issue is the following, where she criticizes:

> The tendency to understand the rejection of scientific objectivity (qua God's eye view on nature) as tantamount to a much stronger (and non sequitur) claim about worldly states of affairs being relative to scientific perspectives.

I take this critical reference to "God's eye view on nature" as analogous to Putnam's "metaphysical realism", and I agree entirely that this is a dogmatic view that should be abandoned. Other analogies between Massimi's views and the late Putnam's views are:

1. The refusal of the antirealist view of truth;
2. The idea of a mind-independent (and perspective-independent) world;
3. A realist semantic tenet about a literal construal of the language of science—i.e., entity realism (which Putnam's based on his own semantic externalism);
4. The idea that (*contra* van Fraassen and the other antirealists and semirealists) accepting a theory implies the belief that the theory is true – i.e., theory realism.

As the late Putnam and Massimi have argued, I don't see any reason to embrace antirealism only because one finds metaphysical realism untenable. A middle-ground view can and should be shaped. The price to pay for that is the abandonment of epistemic monism. At any rate, from the refusal of epistemic monism, Putnam also concluded in favor of non-antinaturalistic ontological pluralism, causal pluralism, and the refusal of the fact-value dichotomy. The main open question here is how monistic one can remain in ontology once one embraces epistemic pluralism—not very much, in my view. In conclusion, if one had to write the history of perspectivism, I would add a section on Putnam's liberal naturalism, as one of the best attempts at overcoming metaphysical realism without abandoning the philosophical realist attitude.

Acknowlewdgements I thank Michela Massimi for her useful comments on a previous version of this paper and David Macarthur for many discussions on the issues treated here. My gratitude toward Hilary Putnam, for our innumerable conversations and his many precious suggestions, is immense.

Bibliography

Block, N. (1997). Anti-reductionism slaps back. *Philosophical Perspectives, 11*, 107–133.

De Caro, M. (2015). Realism, common sense, and science. *The Monist, 98*(1), 197–214.

De Caro, M. (2016). Introduction: Putnam's philosophy and metaphilosophy (pp. 1–18). In Putnam (2016).

De Caro, M., & Macarthur, D. (Eds.). (2004). *Naturalism in question*. Cambridge, MA: Harvard University Press.

De Caro, M., & Macarthur, D. (Eds.) (2010). *Normativity and naturalism*. New York: Columbia University Press.

De Caro, M. & Macarthur, D. (2012). *Hilary Putnam: Artisanal polymath of philosophy* (pp. 1-38). In Putnam (2012c).

Field, H. (1980). *Science without numbers*. New York: Blackwell.

Field, H. (1989). *Realism, mathematics, and modality*. New York: Blackwell.

Field, H. (1992). 'Physicalism', in J. Earman (ed.), *Inference, Explanations, and Other Frustrations: Essays in the Philosophy of Science* (pp. 271-291). Berkeley: University of California Press.

Galilei G. (1632). *Dialogo sopra i due massimi sistemi del mondo*. Engl trans. by S. Drake, *Dialogue Concerning the Two Chief World Systems: Ptolemaic and Copernican*. Berkeley: University of California Press 1967.

Giere, R. (2006). *Scientific perspectivism*. Chicago: Chicago University Press.

Kim, J. (2005). *Physicalism, or something near enough*. Princeton: Princeton University Press.

Maddy, P. (1997). *Naturalism in mathematics*. Oxford: Oxford University Press.

Massimi, M. (2016). Four kinds of perspectival truth. *Philosophy and Phenomenological Research, 96*(2), 342–359.

Massimi, M. (2017). Perspectivism. In J. Saatsi (Ed.), *The Routledge handbook of scientific realism* (pp. 163-175). London: Routledge.

Oppenheim, P., & Putnam, H. (1958). Unity of science as a working hypothesis. In H. Feigl, M. Scriven, & G. Maxwell (Eds.), *Minnesota studies in the philosophy of science*, Vol. II (pp. 3–36). Minneapolis: University of Minnesota Press.

Papineau, D. (1996). *Philosophical naturalism*. Oxford: Blackwell.

Putnam, H. (1973). "Reductionism and the Nature of Psychology." Reprinted in *Words and Life*, ed. by J. Conant (pp. 428-440). Cambridge (MA): Harvard University Press 1994.

Putnam, H. (1975a). Do true assertions correspond to reality? In Putnam 1975b, 70–84.

Putnam, H. (1975b). *Philosophical papers*. Vol. I, *Mathematics, matter and method*. Cambridge: Cambridge University Press.

Putnam, H. (1975c). The meaning of meaning. In Putnam, *Philosophical papers*. Vol. II, *Mind, language and reality* (pp. 215–271). Cambridge: Cambridge University Press.

Putnam, H. (1980). Models and reality. *The Journal of Symbolic Logic, 45*(3), 464–482.

Putnam, H. (1981). *Reason, truth, and history*. Cambridge: Cambridge University Press.

Putnam, H. (1982). The place of facts in a world of values (pp. 142-162). Reprinted in Putnam (1990) 135–141.

Putnam, H. (1990). *Realism with a human face*, ed. by James Conant. Cambridge, MA: Harvard University Press.

Putnam, H. (1999). *The threefold cord. Mind, body, and world*. New York: Columbia University Press.

Putnam, H. (2002). *The collapse of the fact/value dichotomy and other essays*. Cambridge, MA: Harvard University Press.

Putnam, H. (2004). *Ethics without ontology*. Cambridge, MA: Harvard University Press.

Putnam, H. (2005). A philosopher looks at quantum mechanics (again). Reprinted in Putnam 2012c, 126–147.

Putnam, H. (2008). *Jewish philosophy as a guide to life. Rosenzweig, Buber, Levinas, Wittgenstein.* Bloomington: Indiana University Press.

Putnam, H. (2011). The fact/value dichotomy and its critics. Reprinted in Putnam (2012c), 283-298.

Putnam, H. (2012a). Corresponding to reality. In Putnam (2012c), 72–90.

Putnam, H. (2012b). On not writing off scientific realism. Reprinted in Putnam (2012c), 142–162.

Putnam, H. (2012c). *Philosophy in an age of science*, ed. by M. De Caro and D. Macarthur. Cambridge, MA.: Harvard University Press.

Putnam, H. (2015). Naturalism, realism, and normativity. *Journal of the American Philosophical Association, 1*(2), 312–328.

Putnam, H. (2016a). *Naturalism, realism, and normativity*, ed. by M. De Caro. Cambridge, MA: Harvard University Press.

Putnam, H. (2016b). Realism. *Philosophy and Social Criticism, 42*(2), 117–131.

Putnam, H. (2016c). The development of externalist semantics. Reprinted in Putnam (2016a), 199–212.

Putnam, H. (forthcoming). Comment on Charles Travis's 'overflowing bounds' and 'Laudatio'. In W. K. Essler, D. Føllesdal, & M. Frauchiger (Eds.), *Themes from Putnam* (Lauener Library of Analytical Philosophy). Berlin: De Gruyter.

Putnam, H., & Putnam, R. A. (2017). *Pragmatism as a way of life. The lasting legacy of William James and John Dewey*. Cambridge, MA: Harvard University Press.

Rosenberg, A. (2013). Disenchanted naturalism. In B. Bashour & H. Muller (Eds.), *Contemporary philosophical naturalism and its implications* (pp. 17–36). London: Routledge.

Smart, J. J. C. (1978). The content of physicalism. *The Philosophical Quarterly, 28*, 239–241.

Stoljar, D. (2010). *Physicalism*. New York: Routledge.

Chapter 5
Scientific Perspectives, Feminist Standpoints, and Non-Silly Relativism

Natalie Alana Ashton

Abstract Defences of perspectival realism are motivated, in part, by an attempt to find a middle ground between the *realist intuition* that science seems to tell us a true story about the world, and the *Kuhnian intuition* that scientific knowledge is historically and culturally situated. The first intuition pulls us towards a traditional, absolutist scientific picture, and the second towards a relativist one. Thus, perspectival realism can be seen as an attempt to secure situated knowledge without entailing epistemic relativism. A very similar motivation is behind feminist standpoint theory, a view which aims to capture the idea that knowledge is socially situated whilst retaining some kind of absolutism. Elsewhere I argue that the feminist project fails to achieve this balance; its commitment to situated knowledge unavoidably entails epistemic relativism (though of an unproblematic kind), which allows them to achieve all of their feminist goals. In this paper I will explore whether the same arguments apply to perspectival realism. And so I will be asking whether perspectival realism too is committed to an unproblematic kind of relativism, capable of achieving scientific goals; or, whether it succeeds in carving out a third view, between or beyond the relativism/absolutism dichotomy.

Keywords Feminist standpoint theory · Feminist epistemology · Perspectival realism · Relativism

5.1 Introduction

Ronald Giere (2006) presented his perspectival realism as a way to mediate between absolute, realist scientific objectivism on the one hand, and a variety of relativist, anti-realist, social constructivist views, on the other hand. On Giere's view, knowledge is situated, but we can still make meaningful reference to a single 'real' world,

N. A. Ashton (✉)
University of Vienna, Wien, Vienna, Austria

© The Author(s) 2020
A. Crețu, M. Massimi (eds.), *Knowledge from a Human Point of View*,
Synthese Library 416, https://doi.org/10.1007/978-3-030-27041-4_5

independent of any perspective. This sounds very similar, both in content and motivation, to feminist standpoint theory. Standpoint theory is also a view on which knowledge is situated (socially situated, due to factors like a knower's race and gender), and which (at least on some framings) has been presented as a third way between absolute scientific objectivism and relativism about scientific claims.

Elsewhere I have argued that feminist standpoint theory fails at its goal of providing a genuinely distinct 'third way'. On closer inspection, it turns out that (a certain kind of) relativism is essential to the view (Ashton forthcoming); although, unlike most anti-relativist critics of standpoint theory, I don't believe that this diminishes its usefulness in any way. In this paper I apply these arguments I've made about feminist standpoint theory to perspectival realism, to show that perspectival realism (or at least some versions of it) are akin to relativism too.

In Sect. 5.1, I introduce perspectival realism and feminist standpoint theory, and discuss the similarities between them in more detail. In Sect. 5.2, I recount my argument that feminist standpoint theory is a form of relativism. In Sect. 5.3, I apply this argument to perspectival realism, to show that it too is a form of relativism. Giere has expressed concerns about his view being mistaken for, or conflated with, what he calls "silly relativism" (2006, 13)—the view that all perspectives are as good as one another, and thus 'anything goes'. So, in Sect. 5.4, I spend some time specifying what kind of relativism is present in the version of perspectival realism that Giere puts forward. I make clear that it is an unproblematic, *epistemic* relativism, which is not "silly" in the way that Giere warns against.[1] I conclude by considering what perspectival realism and feminist standpoint theories can learn from one another.

Before I start, I am going to run through some preliminary definitions which will be useful to keep in mind for the rest of the paper. Some might seem like very basic terms that most readers will have a handle on already, but these are precisely the terms where nuances can be lost and confusion created if special care is not taken.

First, *realism* and *anti-realism*. These are terms for metaphysical views about 'what there is'. Realists believe that there is an uncomplicatedly mind-independent world, which exists regardless of whether and what we discover about it. Anti-realists believe something other than this: namely, that there is no mind-independent world, or that the world depends in some (important) way on (or is affected by) us and our investigations. One form of anti-realism that Giere talks about is *social constructivism*. Social constructivists believe that (at least some) objects of inquiry are determined (or 'created') by social processes, namely our investigations and our interpretations of the results of these investigations.

The other key terms that I will be using are epistemic terms; these terms apply to claims about what we believe, know, and understand. *Objectivity* is the property (of

---

[1] Some people equate relativism with the silliness Giere has in mind, and to them the idea of non-silly relativism seems like a misnomer – views which aren't silly cannot also be relativism, so if a view is shown not to be silly, it must be something other than relativism (c.f. David Bloor's discussion of "foolishness conditions" (Bloor 2011, 452)). As a precursor to my argument for non-silly (and non-derogatory) relativism, it might be helpful to point out that there are a number of serious, self-described epistemic (or scientific) relativists, including: David Bloor (1976), Lorraine Code (1991), Paul Feyerabend (2010), Martin Kusch (2002), and Richard Rorty (1991).

justification or of knowledge claims) of being independent of social or individual factors. It is often associated with *absolutism*, which is the view that the standards for justification apply universally, regardless of time, place, culture, and so on.[2] These are both typically taken to be in opposition to epistemic *relativism*, which is the view that there are, or can be, different justificatory standards which apply in different times, places, cultures, and so on.

When I talk about relativism I mean *epistemic* relativism, and specifically a view which incorporates three components identified by Martin Kusch (2016, 34–5).[3] The first component is epistemic *dependence*, which says that a belief or claim has an epistemic status (as justified or unjustified) only relative to an epistemic system or practice. This can be seen as a rejection of absolutism, as defined above. Michael Williams (2007, 94) also identifies this component, though his definition only makes explicit reference to systems, not practices. The second component is *plurality*: there are (or have been, or could be) more than one such epistemic system or practice. Plurality is widely accepted as a necessary component of relativism (e.g. Williams 2007, 94; Coliva 2015, 140). The third component, *non-neutral symmetry*, says that there is no neutral (i.e., system-independent) way to evaluate or rank systems or practices (Kusch 2016, 35). It will be worth discussing this component in more detail.

Non-neutral symmetry lines up nicely with the rejection of absolutism that we've seen embodied by dependence. Recall that absolutism is the idea that justification is independent of time, place, culture, and so on. Epistemic dependence rejects this claim, on the ground that justification is dependent on (temporally, geographically, or culturally) contingent systems and practices. Non-neutrality expresses a related point: if justification is system-dependent, then the justification for any evaluation or ranking of a set of systems will be epistemically dependent too. Neutral, or system-independent, rankings are not possible.

As we'll see, discussions of relativism often involve a different version of symmetry, known as equality, or *equal validity*, which says that all epistemic systems and practices are equally correct (Kusch 2016, 35). At first sight this might appear to be almost indistinguishable from non-neutrality—after all, they both seem to say that it is not possible to rank different systems and practices. But on further investigation they are quite different. Neither of them rules out system-rankings altogether, and, unlike non-neutrality, equal validity is incompatible with the rejection of absolutism. Remember, non-neutrality doesn't say that no rankings *at all* are possible—only that system-independent rankings are not. This, as we saw above, is in line with the rejection of absolutism as embodied by dependence. Equal validity doesn't say that no

---

[2] Not to be confused with absolutism in philosophy of science, which is not typically tied (at least directly) to justification. Thanks to Michela Massimi for flagging up this potential source of confusion.

[3] Kusch identifies five essential components of relativism (and several further non-essential ones) in total (2016, 34–6). I've decided to just include three of Kusch's essential components, because there is precedent for this tripartite understanding of epistemic relativism (e.g. Williams 2007; Coliva 2015), and because I believe the other two essential components can be shown to follow from these three.

rankings are possible either, because it itself is a ranking: it says that all systems and practices rank in the same position, and it seems to presume a neutral point (or non-point) of view from which this ranking is made. This directly conflicts with dependence and the rejection of absolutism, but fits naturally within an absolutist picture.

If a view contains the first three components I've mentioned—dependence, plurality, and non-neutral symmetry—then it is, or includes, a form of epistemic relativism. If, instead, (or in addition) it contains equal validity then it is internally inconsistent, and not a form of epistemic relativism. With these preliminaries in place, it's time to properly introduce and compare perspectival realism and feminist standpoint theory.

5.2 Comparing Perspectival Realism and Feminist Standpoint Theory

In this sect. I will outline both perspectival realism and feminist standpoint theory, and then highlight some central similarities between the two views. In doing this I will make clear why I think that the arguments I am about to make about standpoint theory could also be applied to perspectival realism. I will also point to some important differences which should be borne in mind when attempting to draw out connections between perspectival realism and feminist standpoint theory.

Perspectival realism is a view proposed as a *middle way* between objective scientific realism and a variety of anti-realist views, encompassing relativism and constructivism (Giere 2006, 3; Massimi 2018, 164). It is realist to the extent that it allows that there can be a single way the world metaphysically is, but is perspectival in the sense that it says that our knowledge of the world— including our best scientific knowledge—is historically and culturally *situated*, and cannot give a single, complete, objective picture of the world.[4] Giere's comparison between his description of representational scientific models, on the one hand, and maps on the other hand, helps to make this idea clearer (2006, 72–81). Like maps, Giere says that scientific models are partial, limited representations of a particular area or aspect of the world (2006, 72–3). In the same way that multiple, different, and even conflicting, maps can be said to accurately represent the world, multiple incompatible models, can accurately represent the world too (2006, 78–80). This means that we must create, and choose between, models (as we do with maps) based on our interests, and interpret them using existing cultural conventions (2006, 73–4).

Feminist standpoint theories, which comprise one of the main branches of feminist epistemology, are best explained by making reference to three theses.[5] The *situ-*

---

[4] Michela Massimi points out that different authors take different views on whether historical or cultural situatedness takes priority, or whether they are both equally important to perspectival realism (2018, 164).

[5] The other main branches of feminist epistemology are feminist empiricism (e.g., Anderson 1995, Longino 1997) and feminist postmodernism (e.g., Haraway 1988).

ated knowledge thesis, which is common to all branches of feminist epistemology, says that differences in the social situations of inquirers (including things like their race and gender) make for epistemic differences—differences in what they can justifiably believe (Ashton and McKenna forthcoming). Feminist standpoint theory offers two further theses which help to specify the situated knowledge thesis. The *standpoint thesis* says that justification depends on 'socially situated' perspectives. According to this idea, subjects have different 'social locations', or different statuses as socially oppressed or socially privileged. For example, black women occupy very different social locations to white men, and these different social locations come with different experiences, which have the potential to enable different epistemic perspectives. Finally, the *epistemic advantage thesis* says that experiencing social oppression can lead to more, or better, justification. The idea is that subjects who are socially oppressed have distinct experiences, and through critically reflecting on these can turn their perspective into a 'standpoint'—an epistemically privileged perspective from which the nature of relevant social relations is visible. Subjects who are not oppressed do not have these experiences, and as a result are less likely to achieve a standpoint.[6] Different versions of standpoint theory can be distinguished by how they flesh out the details of the epistemic advantage thesis, but all of them take this, and the standpoint thesis, as a starting point. Then it is clear that on standpoint theory, as with perspectival realism, knowledge is situated.

Standpoint theory can also be seen as a *middle way*, between traditional scientific objectivity and what Giere has called "silly relativism" (Giere 2006, 13). Sandra Harding (1991) has explicitly presented it in this way.[7] She rejects the traditional approach to objectivity in science, which she describes an as attempt to eradicate *all* values from science. She says that this is too ambitious, because some values have benefited science (Harding 1991, 144),[8] but also that these ambitions aren't successfully lived up to. When you try to ignore the effect of social factors on science, you end up allowing unconscious, deeply embedded values—including racist and sexist ones—to thrive unmonitored (1991, 143). However, instead of embracing relativism as the alternative to this "weak objectivism", she instead proposes that we conduct science according to the goal of "strong objectivity" (1991, 149–52). This means acknowledging that values have a role in science—i.e., that knowledge is situated—and then paying close attention to them, so that we can be aware of and mitigate their effects.

---

[6] Four important caveats accompany the epistemic advantage thesis. The advantage: doesn't depend on essential categories (e.g., of 'woman'); it isn't a necessary or a sufficient condition on membership of a social group; it isn't 'automatic', but earned through critical reflection; and it can be restricted in scope (Ashton and McKenna forthcoming).

[7] Most standpoint theorists have disavowed relativism (e.g., Medina 2013), and their views are often interpreted as rejections of objectivity due to their commitment to situatedness. Harding is the only author I know of who has explicitly presented her view as a middle way.

[8] C.f. Longino (1997) who argues that some values are necessary when choosing between multiple models or theories that fit the evidence equally well.

To sum up, standpoint theory and perspectival realism have at least two things in common. They both see knowledge as *situated,* by which I mean they both think that knowledge depends on necessarily partial and interested perspectives. And they have each been framed as a *middle way* between traditional scientific objectivism on the one hand, and 'silly', relativism and constructivism on the other. Before moving on, there are however two differences between perspectival realism and standpoint theory that I should flag up. The first is due to Kristina Rolin (2009), who argues that standpoint theories have a distinctive focus on power relations, which Giere's view of scientific models as akin to maps does not capture:

> The map metaphor invites us to see the subject matter of inquiry as a passive object waiting to be represented by someone else. In the map metaphor, scientific knowledge is perspectival in the sense that the mapmakers decide what to represent and to what degree of detail to represent it. In my interpretation of feminist standpoint theory, social-scientific research is perspectival not only in this sense but also in another sense. The term "perspective" refers not only to a social scientist's perspective but also to an informant's perspective, and these two perspectives may or may not meet each other in the inquiry (Rolin 2009, 223).

Whilst Giere emphasises the partiality and contingency of scientific theories, Rolin says that standpoint theories go further by showing that the power relations surrounding the construction and testing of scientific theories can suppress or distort evidence. It's not just that there are multiple ways that scientists can choose to go about 'mapping', but that what they are trying to map can present itself in multiple ways—the terrain can move. The second difference between feminist standpoint theory and perspectival realism is that they come from different sub-disciplines with different sets of terminology. Perspectival realism comes from the philosophy of science, and authors here use the language of "theories", "models" and their "fit with the evidence", whilst feminist standpoint theory comes from epistemology and often makes claims about "belief" and "justification". These terminologies don't translate perfectly, and whilst I don't think this will raise particular problems in this paper, I do think it's important to flag it to avoid confusion.

5.3   Relativism in Feminist Standpoint Theory

In the previous sect. I showed that there are two key similarities between feminist standpoint theory and perspectival realism. The first is that they both involved *situatedness*, and so reject traditional objectivity. The second is that, despite the above, their proponents have been keen to distance themselves from epistemic relativism. Elsewhere I have argued that the anti-relativist arguments that standpoint theorists make don't work, and that feminist standpoint theory is best understood as an illustration of how epistemic relativism can be successful (Ashton, forthcoming). In this paper I'm going to argue that the same is true of perspectival realism, but first I'll briefly explain my argument that standpoint theory is relativist.

Recall that there are three components which I take to be jointly sufficient for epistemic relativism: epistemic dependence, plurality, and non-neutral symmetry.

The first two of these follow quite naturally from the standpoint thesis. Dependence follows because the standpoint thesis says that justification is *dependent* on a perspective. Plurality follows from it because the standpoint thesis says that there is *more than one* of these perspectives. So standpoint theory uncontroversially has two of the three components that are jointly sufficient for relativism. Accordingly, standpoint theorists hoping to disprove the claim that their views are relativist standardly attempt to do so by denying that they endorse the third component of epistemic relativism, symmetry. They do this by pointing to the epistemic advantage thesis, which says that the standpoints of socially oppressed groups provide epistemic benefits (Tanesini 2018). The idea is that the epistemic advantage thesis shows that some standpoints (those of the socially oppressed) are better than others, that standpoint theory allows standpoints to be ranked, and that standpoint theory is therefore not compatible with symmetry. If it isn't compatible with symmetry then it can't include all three of the components which are jointly sufficient for relativism, and so the argument that it is relativist falls apart.

But remember in the Introduction I distinguished between two versions of symmetry: non-neutral symmetry and symmetry based on equal validity. My (*forthcoming*) argument that feminist standpoint theory is relativist turns on this point. Or, rather, on standpoint theorist's failure to recognise this point. Standpoint theorists succeed in showing that the epistemic advantage thesis is incompatible with symmetry based on equal validity. Equal validity says that there is no way at all to rank epistemic systems and that they are all equally correct. The epistemic advantage thesis says that some standpoints are better than others, and so is incompatible with this. But this is all beside the point if, as I have claimed, non-neutrality, rather than equal validity, is what makes a view relativist.[9] Non-neutrality makes a weaker claim than equal validity; it doesn't say that there is no way at all to rank different epistemic systems, it only says that there is no *neutral*, system-independent way to evaluate or rank different epistemic systems. The epistemic advantage thesis isn't incompatible with this version of symmetry, and so its presence isn't sufficient to show that standpoint theory is not a form of epistemic relativism.

The epistemic advantage thesis could be made to be incompatible with non-neutrality (and so with epistemic relativism) if it were interpreted such that the ranking(s) it recommends are system-independent. If the claim that (for example) black feminist standpoints are epistemically superior to the perspectives of white men was supposed to be *absolutely* justified, rather than justified relative to one or more specified epistemic systems, then it would be incompatible with non-neutrality

---

[9] On this definition of relativism, many more accounts of justification and scientific knowledge will turn out to be relativist than we (or even the proponents of these accounts) might initially expect. However, I don't think this is a mysterious result— if a view has almost universally been associated with 'silliness' and incoherence then it shouldn't be surprising to discover that philosophers have been reticent to embrace that label, however unfair the charges of silliness and incoherence are.

and so avoid epistemic relativism.[10] However, I don't think the option is open to standpoint theorists to interpret the epistemic advantage thesis in this way.

The central and defining component of standpoint theory, the *standpoint thesis*, says that justification is dependent on socially-situated perspectives. This means that the justification of a ranking of different systems will be dependent on socially-situated perspectives too, and so standpoint theorists are committed to understanding the epistemic advantage thesis (which is effectively a ranking of different systems) as system-dependent. The epistemic advantage has to be understood as system-dependent if standpoint theorists are going to maintain a consistent view, and so standpoint theory is best understood as a form of epistemic relativism.[11]

5.4 Relativism in Perspectival Realism

In the last sect. I recounted my argument that standpoint theory is a kind of relativism. I argued that, in addition to having the first two components of relativism—dependence and plurality—standpoint theory can also be shown to have the more controversial, third component of relativism, namely non-neutral symmetry. Standpoint theorists can't appeal to an absolutist ranking of standpoints (because of their commitment to dependence) and they (correctly) deny that their view involves symmetry based on equal validity, so the epistemic advantage thesis is best understood in line with non-neutral symmetry. We've already seen that there are significant structural similarities between perspectival realism and standpoint theory, which may suggest that perspectival realism too could contain all three of these components. In this sect. I'll follow through on this, by arguing that Giere's perspectival realism is committed to all three components, and so is a kind of epistemic relativism.

Let's start with the first component. Dependence says that a belief has an epistemic status (as justified or not) only relative to an epistemic system or practice. Giere's perspectivism, like standpoint theory, says that knowledge is situated: scientific claims can only be made using, and evaluated against, scientific models and their associated instruments, methods and interpretive conventions. This means that,

---

[10] Aidan McGlynn raised a similar point in a commentary on an earlier version of this paper: at least some arguments for standpoint theory (those turning on examples of scientific progress made by the introduction of a feminist perspective) seem to justify an absolutist understanding of epistemic advantage. In addition to the response I go on to make above (that this would render standpoint theory inconsistent) I also dispute that McGlynn's is the best way to understand these examples. I believe they can be explained just as well by deploying a relativist understanding of shared progress, which doesn't justify an absolutist understanding of advantage, as they can with an absolute notion of progress. I defend this claim in Ashton (forthcoming).

[11] To be clear, I don't consider this to be a problematic result for standpoint theorists because, as I've argued elsewhere (Ashton forthcoming; Ashton and McKenna forthcoming), the most frequently criticised aspects of relativist feminist epistemologies are much less troubling than they are typically supposed to be.

according to perspectival realism, any scientific knowledge claim we might make is necessarily understood and evaluated relative to an epistemic (scientific) practice, and so its epistemic status is dependent on that practice. Perspectival realism incorporates the dependence component.[12]

The second component is plurality, which says that there are, have been, or could be multiple epistemic systems or practices. This, too, is the case according to perspectival realism, whose proponents point to various different models and practices, such as Aristotelian and Ptolemaic perspectives (models and practices), Copernican and Newtonian ones, and the contemporary perspective of General Relativity (Giere 2006, 94). Of course, these different perspectives share plenty of very general principles, such as principles outlining the most general uses of observation and testimony, but there are other, narrower, principles which are specific to particular perspectives. For example, principles of the form:

"If S uses instrument i, and observes result r, whilst conditions c obtain, then S is justified in believing x."

These are the kinds of principles which scientists from different perspectives disagree over, and which (in combination with more general principles, as well as certain modelling practices) their different beliefs are justified relative to. So the plurality component is also present in perspectival realism.

The final component is symmetry, and specifically non-neutral symmetry, which says that there is no neutral way to rank different epistemic practices. We can see this reflected in Giere's view even more clearly than in feminist standpoint theory. Like standpoint theorists, Giere believes in scientific progress and doesn't judge all perspectives to be on a par. This might be taken to undermine symmetry; where standpoint theorists draw on the epistemic advantage thesis, Giere can appeal to a Kuhnian picture of progress on which perspectives differ in their puzzle-solving power.[13] But, also in common with standpoint theorists, Giere is committed to situated perspectival knowledge. This means that we should expect him to reject absolutism and recognise that the scientific claims he makes, including claims he makes about the ranking of different perspectives, are perspectival too. And he does this; he explicitly rules out the possibility that perspectivism itself is *absolutely* justified and perspective-free, instead acknowledging that it is developed "within the framework of contemporary science", and that his "own claims must be reflexively understood as themselves perspectival" (Giere 2006, 3).

---

[12] To be clear, this needn't be dependence in the sense of *logical* dependence. The knowledge claims in question don't need to be shown to logically follow from a set of propositions or principles that make up the relevant perspective. All that's required is that knowledge claims' epistemic statuses are dependent on a scientific system or practice.

[13] On (at least one reading of) Thomas Kuhn's (1994: 160–73) view, successive scientific paradigms solve more, or more important, puzzles (i.e., they are able to explain more, or more important, disparities between what the paradigm's theories predict and what its experiments appear to show) than their predecessors. Thanks to Michela Massimi for pointing out that this option is available to Giere.

This is precisely the move that I encouraged standpoint theorists to make in the previous section: to take situatedness (or perspectivality) seriously enough to apply it to one's own claims, and to recognise that those claims are situated too. In making this move, Giere accepts non-neutrality. He accepts that there's no neutral way to rank different perspectives, because any such ranking would itself be situated, and dependent on a scientific perspective. Like feminist standpoint theory, Giere's perspectival realism incorporates dependence, plurality, and non-neutral symmetry, and so is a form of epistemic relativism. However, unlike standpoint theory, perspectival realism's commitment to non-neutrality (at least in the work of Giere) is made much more explicit.

5.5   Non-Neutral and Non-silly Epistemic Relativism

I've just argued that perspectival realism, like standpoint theory, is a form of relativism. Relativism has a negative reputation and describing a view as relativist is typically seen (and often also intended) as being critical of that view. This is not my intention however. I think the form of relativism found in perspectival realism is unproblematic. In this section I will specify exactly what sort of relativism is present in perspectival realism, making reference to Massimi's account which builds on Giere's work, and—perhaps more importantly—I'll show what kind of relativism is not present in perspectival realism. The relativism present in perspectival realism is one that perspectival realists should embrace. The first specification I want to make is that perspectival realism is a form of epistemic, not metaphysical, relativism. Taking a closer look at Giere's taxonomy of different views will help to make this point more clearly.

Giere frames his perspectival realism as an alternative to existing views: scientific realism, and social constructivism. It will help to separate what each of these views have to say about metaphysics (how the world is) and epistemology (what we can legitimately say about the world).[14] Giere seems to categorise scientific objectivism as *metaphysically* realist (i.e., there is a single way the world is) and *epistemically* objectivist (i.e., there is a single correct description of that world that science strives to obtain, see 2006, 4–5). Massimi's perspectival realism takes the first (metaphysical) component whilst rejecting the second (epistemic) one. Giere then explains that social constructivism is *metaphysically* anti-realist (i.e., it denies there's a single way the world is),[15] and *epistemically* perspectival (i.e., it says there are multiple legitimate ways of describing and understanding the word).[16] Massimi's

---

[14] This taxonomy almost certainly oversimplifies matters, but it will help to get a rough idea of the logical space.

[15] Giere (2006, 7) cites the strongest version of social constructivism, according to which we construct the world by constructing different sets of facts Latour and Woolgar (1979).

[16] Here Giere (2006, 7–8) cites less radical versions of social constructivism: Bloor (1976); Collins (1981); and Shapin (1975, 1979).

perspectival realism takes the latter (epistemic) component of this and rejects the former (metaphysical) one. The table below summarises this.

	Scientific Realism	Perspectival Realism	Social Constructivism
Metaphysics	Realist	Realist	Anti-realist
Epistemology	Objectivist	Perspectivist	Perspectivist

So, on this understanding, perspectival realism combines the metaphysical view that there is a single way the world is with the epistemic view that there are multiple, legitimate ways that we can view or conceptualise the world. Its perspectivism is only present in the epistemic domain, making it an *epistemic* relativism. This means that perspectival realism avoids any worries associated with metaphysical or alethic relativism. I take this to be Massimi's (2017) point when she considers the compatibility of perspectivism and realism. She emphasises the distinction between metaphysical and epistemic relativism, saying that the problem in the past has been people conflating the (epistemic/scientific) rejection of *objectivity* with relativism about *truth* (2017, 170). Even with this specification in place, worries about *epistemic* relativism might still remain. Giere considers two in particular, which I therefore need to address. Both of these turn out to be unproblematic because of the second specification that I will make: perspectival realism is a form of non-neutral relativism, rather than an equality-based relativism.

The first worry is that a relativist version of perspectival realism would run afoul of 'the reflexive question'. Giere first mentions this question in connection with social constructivist theories. It asks whether or not views' conclusions about contingency apply to themselves. If they don't, then holders of that view (in that case constructivism, in this case perspectival realism) grant their own conclusions exceptional objectivity—a surprising position, which requires further justification. And if they do, then holders of the view are forced to admit that their own conclusions are merely contingently (or relatively) justified (Giere 2006, 11).

Giere seems to consider this question to be problematic for social constructivists. He says that they have a "way out", but cautions that it means aligning themselves with radical critics of science (2006, 11), whereas he claims that "no such problems" arise for the perspectival realist (Giere 2006, 95). However, my suggestion in the previous section, that perspectival realists embrace non-neutrality, might be thought to introduce these problems. It seems to commit perspectival realists to something like the second answer to the reflexive question, forcing them to admit that their own claims are 'merely' relatively justified.

Whilst I do think that perspectival realists are committed to the second response to the reflexive question—i.e., to saying that their conclusions themselves are perspectival, and so are contingently or relatively justified—I don't think that this should be considered costly.

The apparent cost of relative justification is that it is 'mere', or a somehow lesser form of, justification. But lesser than what? This worry presumes that there is some other, bigger, better form of justification which relative justification falls short of.

This is an absolutist view. On any relativist view (including perspectival realism) which denies both the existence and possibility of absolute justification, contingent, relative, perspectival justification is the best and most legitimate form of justification available. So if perspectival realists acknowledge and accept the relativism in their view, then they can endorse the second answer to the reflexive question without undermining themselves, and without casting aspersions on science.

Massimi's (2018) account shows us one possible way to flesh out this idea. She proposes that scientific perspectives play a "double role", where they function both as a *context of use*, which fixes its own standards of performance adequacy, and as a *context of assessment*, from which other perspectives can be evaluated (according to how well they meet the standards of performance adequacy which they have set 2018: 353–7). On this view, perspectives aren't the sole measure of their own success, because other perspectives functioning as a standard of assessment are important too, and so worries about lesser or redundant forms of justification are avoided. But no absolute, or perspective-independent, evaluation is required either, and so both non-neutrality and reflexivity are present.[17]

The second worry is that epistemic relativism "can be pushed to the absurd extreme that every perspective is regarded as good as any other" (Giere 2006, 13). This view is sometimes summarised using the phrase 'anything goes' (because all options are equally epistemically permissible). Giere denies that this conclusion follows from his view, making a distinction between perspectivism and "silly relativism". I agree with the letter of this claim, but perhaps not the spirit.[18] Regardless, I will explain why the 'anything goes' conclusion doesn't follow from non-neutrality in particular, and why relativism based on non-neutrality is therefore non-silly.

It's easy to see why this worry arises. The 'anything goes' conclusion can be derived from equal validity, which we've seen is commonly (though mistakenly) associated with relativism. Equal validity says that there is no way to rank epistemic systems whatsoever, and so (from an absolute, independent non-perspective) they are all as good as each other. On this sort of view, any perspective is legitimate, and so: anything goes. But as equal validity is not a component of perspectival realism, this source of the worry is not a concern for us.

Perspectival realism is only committed to non-neutrality. Non-neutrality comes from the idea that all justification is dependent, or perspectival, and so views incorporating non-neutrality aren't able to state *absolutely* that one perspective is better than another. This might seem like another reason to worry about the anything goes conclusion besides equal validity. However, it's not. It is possible to say on non-

---

[17] Massimi denies that her account is relativist, framing it as an alternative to both relativism and scientific objectivism. Whilst I agree that her account (2018) is not a form of alethic relativism (as discussed above), I believe it is a version of (non-silly) epistemic relativism, i.e., it incorporates dependence, plurality, and non-neutrality. In conversation Massimi has suggested that her view avoids non-neutrality (and so relativism) because it is able to account for scientific progress, however I have argued (forthcoming) that scientific progress and non-neutrality are compatible.

[18] It's not clear to me whether Giere thinks that all relativism is silly, or whether he accepts (as I do) that perspectival realism is a form of non-silly relativism and was merely distinguishing this from a subset of relativist views which are silly.

neutral relativism that some perspectives are better or worse *relative to a particular perspective*. In other words, it is possible to provide a non-neutral ranking of different epistemic systems. As long as one is taking a particular perspective—and according to perspectival realists this is always the case—it is possible to rank and evaluate epistemic systems, and so it is not the case that 'anything goes'.

I don't expect the promise of this relativized ranking to be enough to satisfy everyone. I've found there to be a common and strong absolutist intuition that some views are *just better* (or worse) than others, in an unrestricted and unmitigated sense, and so if perspectival realism can't account for this it might leave some people disappointed. I am sympathetic to this intuition, but it is an absolutist intuition, incompatible with views like perspectival realism and standpoint theory which are based on the situatedness of knowledge. Perspectival realists cannot consistently allow scientists to say that practices other than their own are *absolutely* unjustified. What they can do is show that non-neutral, perspectival rankings satisfy the criteria for a useful and desirable philosophy of science. This is what Giere and Massimi already claim to do: they argue that perspectival realism reconciles a useful scientific realism with a plausible understanding of knowledge from a human point of view. Acknowledging that the relativist label applies to this view clarifies, rather than undermines, this work.

5.6 Conclusion

I began this paper by pointing to some similarities between perspectival realism and feminist standpoint theory. I said that they both see knowledge as situated, and that they have each been framed as a middle way between scientific objectivism on the one hand, and anti-realist or relativist views on the other. I then argued for a further similarity. The standpoint thesis which is central to feminist standpoint theory commits them to dependency, and ultimately to a kind of relativism, and the perspectivism in perspectival realism does the same for their view. I have now clarified that the version of relativism present in perspectival realism is unproblematic. It is an epistemic relativism, which avoids the complications of metaphysical versions of relativism, and a non-neutral, non-silly relativism, which avoids the 'anything goes' conclusion. In closing, I will suggest two places where perspectival realism and feminist standpoint theory can learn from each other.

First, I think that perspectival realism has an opportunity to learn from feminist standpoint theory by expanding the range of factors that constitute a perspective. I haven't argued for this here, but feminist standpoint theorists have argued that decisions about which instruments to use, which methods to deploy, and which interpretative conventions to rely on aren't just affected by factors that scientists are consciously aware of. They argue that they (and other aspects of the scientific process) are also affected by social factors like race and gender, that scientists might not be consciously aware of. Feminist standpoint theorists have had a hard time convincing most 'mainstream' epistemologists that this is the case, I think largely because philosophers who are committed to absolutism and objectivity are primed to see views which don't have these commitments as attacks on science and its goals, rather than suggestions for how to achieve those goals more effectively. I suspect that perspectival realists won't have this problem, as their project is accepted as part of 'mainstream' philosophy of science and will thus be more amenable to the kinds of arguments that standpoint theorists make.

Second, I think that perspectival realists have something to teach standpoint theorists, namely that to be consistent they need to embrace reflexive perspectivality (or situatedness). Giere readily accepts that his own claims are as situated as those that he is attempting to theorise about, but standpoint theorists have been reticent to do this. This is a problem, because if they don't recognise that their own claims (in particular the epistemic advantage thesis) are situated too, then they are not fully internalising the situated knowledge thesis and are allowing a tension to creep into their view. Standpoint theorist's reticence to do this has, as far as I can tell, has been motivated by the goal of avoiding epistemic relativism, in an attempt to be taken seriously by 'mainstream' philosophers of science. My hope is that seeing this move made by perspectival realists, who are more mainstream, will ease these concerns and help feminist standpoint theorists too to embrace (non-silly) epistemic relativism.

Acknowledgments This paper was assisted by funding from the ERC Advanced Grant Project "The Emergence of Relativism" (Grant No. 339382), and benefited from very helpful discussion at the "Knowledge from a Human Point of View" workshop in Edinburgh. I'd particularly like to thank Aidan McGlynn for his commentary on the version presented there, and Michela Massimi, Liam Kofi Bright, Katharina Sodoma, Katherina Kinzel, and Martin Kusch for invaluable discussion outside of the workshop.

Bibliography

Anderson, E. (1995). Feminist epistemology: An interpretation and a defense. *Hypatia, 10*(3), 50–84. https://doi.org/10.1111/j.1527-2001.1995.tb00737.x.

Ashton, N.A. (forthcoming). Relativising epistemic advantage. In *The Routledge handbook of philosophy of relativism.*

Ashton, N.A. (forthcoming). Evidence, relativism and Progress in feminist standpoint theory. In *The Routledge handbook of evidence.*

Ashton, N.A., & McKenna R. (forthcoming 2018). *Situating feminist epistemology.* Episteme.

Bloor, D. (1976). *Knowledge and social imagery.* London: Routledge.

Bloor, D. (2011). Relativism and the sociology of knowledge. In S. Hales (Ed.), *A companion to relativism.* Malden: Wiley-Blackwell.

Code, L. (1991). *What can she know? Feminist theory and the construction of knowledge.* Ithaca: Cornell University Press.

Coliva, A. (2015). *Extended rationality: A hinge epistemology.* Palgrave Macmillan.

Collins, H. M. (1981). Stages in the empirical Programme of relativism. *Social Studies of Science, 11*, 3–10.

Feyerabend, P. (1975/2010). *Against method* (4th ed.). London: Verso.

Giere, R. N. (2006). *Scientific Perspectivism.* Chicago: University of Chicago Press.

Haraway, D. (1988). Situated knowledges: The science question in feminism and the privilege of partial perspective. *Feminist Studies, 14*(3), 575–599.

Harding, S. (1991). *Whose science? Whose knowledge?: Thinking from Women's lives*. Ithaca: Cornell University Press.

Kuhn, T. S. (1962/1994). *The structure of scientific revolutions*. Chicago: University of Chicago Press.

Kusch, M. (2002). *Knowledge by agreement, the programme of communitarian epistemology*. Oxford: Oxford University Press.

Kusch, M. (2016). Wittgenstein's on certainty and relativism. In *Analytic and continental philosophy methods and perspectives. Proceedings of the 37th international Wittgenstein symposium* (Vol. 23). Berlin, Boston: De Gruyter.

Latour, B., & Woolgar, S. (1979). *Laboratory life: The social construction of scientific facts*. Beverly Hills: Sage Publications.

Longino, H. E. (1997). Feminist epistemology as a local epistemology. *Aristotelian Society Supplementary, 71*, 19–36.

Massimi, M. (2017). Perspectivism. In J. Saatsi (Ed.), *The Routledge handbook of scientific realism* (pp. 164–175). New York: Routledge.

Massimi, M. (2018). Four kinds of perspectival truth. *Philosophy and Phenomenological Research, 96*, 342–359.

Medina, J. (2013). *The epistemology of resistance: Gender and racial oppression, epistemic injustice, and the social imagination*. Oxford: Oxford University Press.

Rolin, K. (2009). Standpoint theory as a methodology for the study of power relations. *Hypatia, 24*(4), 218–226.

Rorty, R. (1991). *Objectivity, relativism, and truth* (Philosophical Papers, v. 1). Cambridge: Cambridge University Press.

Shapin, S. (1975). Phrenological knowledge and the social structure of early nineteenth-century Edinburgh. *Annals of Science, 32*, 219–243.

Shapin, S. (1979). The politics of observation: Cerebral anatomy and social interests in the Edinburgh phrenology disputes. *The Sociological Review, 27*, 139–178.

Tanesini, A. (forthcoming 2018). Standpoint theory then and now. In *Routledge handbook of social epistemology*.

Williams, M. (2007). Why (Wittgensteinian) contextualism is not relativism. *Episteme, 4*(1), 93–114.

Chapter 6
Perspectives, Questions, and Epistemic Value

Kareem Khalifa and Jared Millson

Abstract Many epistemologists endorse true-belief monism, the thesis that only true beliefs are of fundamental epistemic value. However, this view faces formidable counterexamples. In response to these challenges, we alter the letter, but not the spirit, of true-belief monism. We dub the resulting view "inquisitive truth monism", which holds that only true answers to relevant questions are of fundamental epistemic value. Which questions are relevant is a function of an inquirer's perspective, which is characterized by his/her interests, social role, and background assumptions. Using examples of several different scientific practices, we argue that inquisitive truth monism outperforms true-belief monism.

Keywords Perspectivism · Questions · Epistemic value · Modeling · Idealization · Truth · Belief

6.1 Introduction

We place a great deal of value on inquiry's deliverances. From finding one's keys to discovering the Higgs boson, our quest for knowledge and other cognitive goods is a staple of the human condition. For this reason, epistemologists have increasingly attended to questions of *epistemic value*.[1] A central debate is whether everything that is epistemically valuable bottoms out in true belief's value. Call those who take true belief as the fountainhead of all other epistemic value *true-belief monists*.

True-belief monists are hard-pressed to account for science's many facets. For instance, their narrow edifice seems ill-positioned to explain the complexity of epistemically valuable undertakings characterizing the scientific endeavor. If such

---

[1] Pritchard, Turri, and Carter (2018) review the relevant literature.

K. Khalifa (✉)
Middlebury College, Middlebury, VT, USA
e-mail: kkhalifa@middlebury.edu

J. Millson
Agnes Scott College, Decatur, GA, USA

© The Author(s) 2020   87
A. Crețu, M. Massimi (eds.), *Knowledge from a Human Point of View*,
Synthese Library 416, https://doi.org/10.1007/978-3-030-27041-4_6

worries are well-placed, then true-belief monism is insufficient to account for the epistemic value animating scientific activity. However, we shall focus on whether true-belief monism is necessary for accounting for science's epistemic goods. Several scientific practices involve idealizations (which are false), rely heavily on models (which are frequently, if not fundamentally, non-propositional), and traffic in public, intersubjective representations (and are thereby non-doxastic). For these reasons, one might think that either a pluralistic approach to epistemic value (Pedersen 2017) or a monism that prizes a more flexible epistemic status—such as understanding (Elgin 2017)—should supplant true-belief monism.

In this paper, we have two goals. First, we will clarify scientific challenges to true-belief monism's claim to provide necessary conditions on epistemic value. Second, we will argue that while the letter of true-belief monism is not defensible, a near-neighbor that retains its spirit is. To that end, Sect. 6.2 offers a more precise definition of true-belief monism. Section 6.3 presents our successor position, *inquisitive truth monism*, according to which the ultimate bearers of epistemic value are true answers to relevant questions. Sections 6.4 and 6.5 then show how inquisitive truth monism outperforms true-belief monism in accounting for science's epistemically valuable facets. Finally, one of our boldest departures from true-belief monism is our reliance on *perspectives* to determine which questions are relevant to a given inquirer. For this reason, Sect. 6.6 defends our perspectivism against several potential objections.

6.2 True-Belief Monism

To set the stage, we define true-belief monism as follows:

True-Belief Monism (TBM): For all *x*, *x* is epistemically valuable only if:

1. *x* is a true belief, or
2. *x* is a means to acquiring true beliefs, or
3. *x*'s epistemically valuable components are either true beliefs or a means to acquiring true beliefs.[2]

True-belief monists include Ahlstrom-Vij (2013), David (2001), Goldman (1999), Olsson (2007), and Sosa (2003). Sometimes, this doctrine is put in terms of true belief being the only "fundamental epistemic value" (Ahlstrom-Vij and Grimm 2013). So far as we can tell, TBM's three conditions capture this sense of fundamentality. Corresponding to these conditions are three strategies for true-belief monists to account for epistemically valuable states of affairs. First, per TBM1, whenever something is deemed epistemically valuable because it is a true belief, we will call this the *basic TBM strategy*. Second, whenever true-belief monists account for

---

[2] Because we only discuss whether true-belief monism imposes *necessary* constraints on epistemic value, TBM is not formulated as a biconditional; ditto for inquisitive truth monism.

something's epistemic value by way of TBM2, we will call this *TBM's instrumental strategy*.[3] In general, a state of affairs x is of *instrumental epistemic value* if it is a means to some other epistemically valuable state of affairs y. It is of *mere* instrumental epistemic value if this exhuasts its epistemic value. True-belief monists hold that anything of mere instrumental epistemic value must ultimately be a means to acquiring true beliefs. Paradigmatically, many true-belief monists take justification to be of mere instrumental epistemic value.

Finally, per TBM3, certain epistemic statuses might have derivative epistemic value because their components submit to the basic and instrumental TBM strategies. Call the employment of this claim *TBM's componential strategy*. For instance, true-belief monists might argue that knowledge is of non-fundamental epistemic value because its only non-instrumental epistemic value is true belief, and its remaining components (e.g., justification, Gettier-resistance) are of mere instrumental epistemic value, i.e., are epistemically valuable only insofar as they are an effective means for acquiring true beliefs. Of course, this isn't the only way that TBM can account for knowledge's epistemic value; we use it simply for purposes of illustration.

6.3 Inquisitive Truth Monism

As already mentioned, we have sympathies with true-belief monism, but prefer a kindred position. To motivate this alternative, consider a metaepistemological question: What counts as evidence when adjudicating between competing accounts of epistemic value? Unlike more venerable philosophical topics, e.g., knowledge, truth, goodness, beauty, etc., the term "epistemic value" does not appear in ordinary talk, and our folkways do not make hard and fast distinctions between epistemic and other kinds of value. So, it's unclear which intuitions, practices, etc. matter when claiming that something is or is not epistemically valuable. This makes it too convenient to discard hard cases as lacking in epistemic value (e.g., as merely pragmatic). Furthermore, like many human endeavors, links of causation and covariance are noisy and underdetermined by armchair observations. Consequently, it is too easy to provide just-so stories about how something could be a means to one's favored epistemic good.

Combined, these metaepistemological worries paint an unflattering portrait of discussions concerning epistemic value. For instance, true-belief monists can run their three strategies in fairly unconstrained, *ad hoc* ways, and then discard the remaining cases as non-epistemic on fairly unprincipled grounds. Fortunately, epistemic value theorists in general, and true-belief monists in particular, broadly agree on one pre-theoretic consideration that fixes the reference of "epistemic value"

---

[3] Throughout, we treat instrumental epistemic value as the main kind of derivative epistemic value. Our arguments would not be adversely affected if we countenanced other forms of derivative epistemic value, as Berker (2013) does.

(Ahlstrom-Vij 2013; Lynch 2004; Sosa 2003). Ahlstrom-Vij and Grimm (2013, 330) nicely summarize this idea:

> ... epistemic value is a function of the goals of inquiry, where 'inquiry' refers to the range of inquisitive practices concerned with the posing and answering of questions ... the goals of inquiry determine which activities, states, processes, practices, and so on are epistemically valuable.

In other words, the foremost "theory-neutral" way of determining whether something is epistemically valuable appeals to the goals underlying practices of asking and answering questions. Call this the *Intuition About Inquiry*. We will recruit this platitude to adjudicate between competing accounts of fundamental epistemic value via three interventions.

First, the Intuition About Inquiry suggests that closer attention to the semantics and pragmatics of questions and answers (Belnap and Steel 1976; Groenendijk and Stokhof 1984; Hamblin 1958; Millson 2014; Wiśniewski 1995) mitigates the aforementioend metaepistemological worry. This research clarifies these inquisitive practices and thereby promises to also clarify what is of epistemic value. Space being limited, we can only hint at the possibilities of this here. For present purposes, we adopt what is known as the 'set-of-answers' methodology in the logic of questions and answers, i.e., erotetic logic (Wiśniewski 2013). Following this approach, we treat interrogatives' content as the set of their possible answers. We take possible answers to be propositions and thus conceive of questions as sets of propositions. This allows us to import various relations among questions and statements studied by Inferential Erotetic Logic (IEL) (Wiśniewski 1995, 2013); most importantly, that of erotetic implication (see below). For this essay's purposes, 'answer' is shorthand for 'possible answer'. We rest on readers' intuitions about what counts as a possible answer to a question.

Second, the Intuition About Inquiry suggests that examining scientific inquiries—where inquirers are especially explicit about the questions they are asking and how they are coming to their answers—will further mitigate our metaepistemological worry. Sections 6.4 and 6.5 embrace this suggestion in full. Of course, scientific inquiries are not the *only* inquiries worth consulting. Rather, we merely suggest that scientific inquiries are especially useful in adjudicating between competing accounts of epistemic value.

Third, and most importantly, the Intuition About Inquiry hints at an alternative to TBM. It is a platitude that the goal of asking a question is answering it correctly. This suggests the following:

Inquisitive Truth Monism (ITM): For all x, x is epistemically valuable only if:

1. x is a true answer to a relevant question, or
2. x is a means to acquiring true answers to relevant questions, or
3. x's epistemically valuable components are either true answers to relevant questions or a means to acquiring true answers to relevant questions.

Like TBM, ITM admits of basic, instrumental, and componential strategies and prizes truth. However, TBM valorizes true *beliefs*, while ITM valorizes true *answers to relevant questions*. As we shall argue below, belief is often ancillary to scientific inquiry. Conversely, true beliefs that answer no relevant questions fail to fulfill scientific inquiry's goals. As such, ITM enjoys certain explanatory advantages over TBM.

According to our brand of ITM, *perspectives* determine whether a question is relevant or not. Perspectives are combinations of an agent's personal interests, social roles, and background assumptions.[4] We discuss each in turn. First, questions may be relevant because of inquirers' *personal interests*.[5] Interests in questions' answers may serve some practical end, e.g., "Where is the rake?" Other questions might be sparked by interests that are less practical, e.g., mere curiosity.

Second, questions may be relevant because inquirers have *role-responsibilities* to answer them.[6] Here, we adopt Hart's (1968, 212) account of role-responsibilities:

> … whenever a person occupies a distinctive place or office in a social organization, to which specific duties are attached to provide for the welfare of others or to advance in some specific way the aims or purposes of the organization, he is properly said to be responsible for the performance of these duties, or for performing what is necessary to fulfill them.

As a simple illustration, John would be within his rights to ask his mechanic, Jane, why his car screeches when he makes left turns. Furthermore, Jane would be failing to fulfill her professional responsibilities if she refused to explain this to John. As such, her role-responsibility includes answering the question, "Why does John's car screech whenever he turns left?". The question is thereby relevant to Jane, even if she is not interested in the answer.

Finally, questions may be relevant in a 'derivative' sense, so to speak. For instance, suppose that the following question is relevant: "Is every member of the Edinburgh philosophy faculty Scottish?". Then each of the following questions are also relevant: "Is Michela Massimi Scottish?", "Is Nick Treanor Scottish?", etc. In IEL's parlance, the first "superquestion" *erotetically implies* the other "subquestions". So, relevant superquestions can "transmit" their relevance to subquestions via erotetic implication. Much like role-responsibilities, erotetic implications place inquirers on the hook for questions that they might not be interested in answering— or even aware of.

Importantly, erotetic implication can be far more sophisticted than the example above suggests. Erotetic implication frequently behaves nonmonotonically, which pairs naturally with it being sensitive to an inquirer's background assumptions (Millson 2019). For instance, suppose "Where is the rake?" is a relevant question

---

[4] Some might equate our notion of perspective with a "social-epistemic context". We have no quarrel with this way of speaking. However, because different scholars ascribe different meanings to the phrase "social-epistemic context", we seek to mitigate misinterpretation by construing perspectives in terms of personal interests, social roles, and background assumptions.

[5] Hereafter, the term "interest" is shorthand for "personal interest", and is meant to denote any interests that would not be "professional" or "social" interests falling out of role-responsibilities.

[6] Elsewhere (Millson and Khalifa in press) we argue for this position at length, but only with respect to why-questions. Nothing seems to prevent our arguments from carrying over to other kinds of questions, as we assume here.

and that the speaker remembers that her partner tends to leave the rake outside when in a rush. This question then erotetically implies the subquestion, "Was my partner in a rush?". Since the background assumptions include the speaker's memories, the inference can be overturned if it turns out that the speaker's memory is unreliable. Importantly, we do not assume that background assumptions need to be true in order to generate relevant questions. We defend this position in Sect. 6.6.

To summarize thus far, we are proposing ITM in lieu of TBM. Its major difference is in replacing true beliefs with true answers to relevant questions as the locus of epistemic value. We are perspectivists about questions' relevance, where agents' perspectives include interests, social roles, and background assumptions. A question is relevant to an inquirer if she is interested in its answer, has a role-responsibility to answer it, or it is a subquestion that is erotetically implied by a superquestion (and her background assumptions) that is already relevant because of these interests and role-responsibilities. Thus, ITM implies that whether or not a truth is epistemically valuable depends on one's perspective.

6.4   Characterizing Inquiry's Goals

We begin by arguing that inquisitive truth monism (ITM) better describes scientific inquiry's goals than true-belief monism (TBM). This argument rests on two claims.[7] First, true beliefs that fail to answer relevant questions are not among inquiry's goals. Second, some true answers to relevant questions need not be believed in order to serve as these goals. Consequently, ITM more faithfully describes inquiry's goals than does TBM.

Begin with the first claim. According to the Intuition About Inquiry, we should consult our practices of asking and answering questions to determine what is epistemically valuable. True beliefs that answer no relevant questions seem to clash with this intuition and common sense, for belief in any true *non sequitur* would thereby function as a goal of inquiry. For instance, suppose that someone's only true belief about the ideal gas law is that Clapeyron first formulated it in 1834, but she seeks to answer the question, "Why does the ideal gas law hold at low pressure and high temperature?". Clearly, she has not met this inquiry's goal despite having a true belief. Hence, absent ITM's constraints, TBM fails to characterize inquiry's goals. Moreover, note that the concept of relevance is precisely what is perspectival about ITM. Hence, ITM's perspectivism makes all the difference in its superior coverage of inquiry's goals.

Turn now to the second claim: that having a true answer to a relevant question without believing it can sometimes result in a successful inquiry. TBM denies that such inquiries are successful, while ITM strongly suggests that they are.[8] However,

---

[7] We assume a fairly intuitive notion of belief. It exceeds the scope of this paper to see if conceptions of belief that eschew these intuitions (e.g., Hieronymi 2008) compete or complement ITM.

[8] ITM only suggests, but does not entail, this, because we are silent regarding epistemic value's sufficiency-conditions.

some other cognitive attitudes toward a true answer appear to do just as well as belief in properly settling a scientific inquiry. ITM suggests that scientists who rightly believe an answer have nothing more epistemically valuable than scientists who, e.g., merely accept it. Here, we follow Cohen (1992) in defining *acceptance* as the adopting of a policy of including a claim as a premise in deciding what to do or think within a given context, where (as we see it) the most important things to "do or think" involve answering the relevant questions. While we focus on acceptance, there is no reason these points could not be extended to other attitudes. For instance, Buchak (2014) argues that credences are distinct from beliefs. If arguments such as hers are sound, then our remarks about acceptance carry over to credence. More generally, we might think of 'having' true answers to relevant questions as mulitply realized by different kinds of mental states.

Leaving room for acceptance without belief accords well with scientific practice. Some notable scientists, including Ernst Mach, Karl Pearson, the young Einstein, Niels Bohr, and Milton Friedman, did not appear to believe some of their most notable scientific discoveries, but can plausibly be interpreted as accepting those results. Indeed, some have suggested that scientific inquiry *typically* involves acceptance rather than belief (Kukla 2015). By contrast, TBM implausibly implies that scientists who accept without believing answers possess nothing of epistemic value.

Importantly, our discussion of acceptance is just the tip of the iceberg: ITM is even more inclusive about the vehicle by which a true answer is delivered. Many scientific representations are public and intersubjective. Hence, they are not individual mental states. While this claim seems intuitive, it gains further traction when we consider that aside from recognition and esteem, scientific innovators have little in the way of property rights with respect to their discoveries (Merton 1942/1973). Similarly, paradigmatic uses of the term 'scientific knowledge' frequently refer to epistemic statuses that do not supervene on individual mental states (Bird 2010, Magnus 2013). Thus, if beliefs are individual mental states, then scientific inquiry frequently aims for something else.

In other words, scientific representations may be of a more public flavor than TBM would allow. Non-mental representations (such as declarative sentences) or speech-acts that need not express belief (such as public announcements) can also express answers to relevant questions, and our view takes them to be just as epistemically worthwhile as true beliefs. These different vehicles—sentences, public announcements, collective acceptance, and other kinds of "public answers"—show that answers to relevant questions need not wear the cloak of belief. Thus, unlike TBM, ITM has no difficulty in accommodating inquiries that have these intersubjective representations as their goal.

Taking stock, having true answers to relevant questions without belief can sometimes result in successful inquiry. Answers can be accepted without being believed, or can be public representations without expressing beliefs. Furthermore, the vehicle by which true answers are delivered is naturally thought to depend on one's interests and social roles, i.e., one's perspective. Hence, perspectivism once again is ITM's lynchpin.

Yet, true-belief monists may resist our arguments about public answers in at least three ways. We anticipate and rebut these objections in what follows. First, true-belief monists may counter that, at root, scientists aim to have true beliefs, and that public answers are merely pragmatic side-effects of this true belief. Let *attitudinal ecumenicalism* denote ITM's ability to accommodate both doxastic and non-doxastic attitudes as goals of inquiry. Then this defense of TBM merely presupposes the falsity of attitudinal ecumenicalism without argument; it does not preclude the possibility that scientists sometimes aim to *accept* true answers. So, even if public answers are mere side-effects, it is not clear that they are mere side-effects of *beliefs*. Moreover, on this defense, TBM concedes that beliefs are causally efficacious in producing public answers. However, they then need to fend off an alternative interpretation: beliefs are a mere means to public answers. For instance, providing an answer as a public representation (e.g., a declarative sentence) frequently requires belief simply because one's role-responsibilities require sincere communication. In such inquiries, belief's value might be exhausted by this sincerity requirement.

Second, TBM's adherents may claim that inquiry sometimes aims at collective beliefs, and public answers ought to be collectively believed. After all, TBM says nothing about whether beliefs are restricted to individuals. However, this once again fails to counter attitudinal ecumenicalism, for belief is not the only attitude amenable to collectivization: collective acceptance, collective credence, etc. may sometimes serve as the aim of inquiry. Indeed, some compelling arguments suggest that collective acceptance is more apt to ascribe to groups than is collective belief (Wray 2001). Clearly, such arguments dovetail with our previous remarks.

Finally, TBM's defenders may treat public answers as a means to getting many inquirers to have true beliefs. Yet again, this does not address attitudinal ecumenicalism. In other words, purveyors of this defense owe some explanation for why public answers are a means to *beliefs* rather than to any number of other cognitive attitudes one may adopt towards true answers to relevant questions. Moreover, it suggests that scientists who had true but unpopular answers during their time—e.g., Nicolas Copernicus, Ignaz Semmelweis, and Rosalind Franklin—only succeeded in their inquiries when their respective scientific communities came around. By contrast, ITM more plausibly suggests that they succeeded when they correctly answered their inquiries' main questions, regardless of how slow their peers' uptake may have been.[9]

To summarize, true beliefs that fail to answer relevant questions are not goals of inquiry, since they would allow *non sequiturs* to serve as such goals. Conversely, some true answers to relevant questions—including accepted and public answers—need not be believed in order to serve as these goals. Consequently, ITM more faithfully describes inquiry's goals than does TBM.

---

[9]As before, ITM can only *suggest* this because we are not arguing for its sufficiency.

6.5 Scientific Practice

True-belief monism (TBM) and inquisitive truth monism (ITM) differ in subtle ways. In the prevailing literature on epistemic value, differences such as these are typically adjudicated by comparing competing views' accordance with relatively mundane epistemic practices, sometimes spruced up through the art of thought experiment. However, we shall take a different tack, as many of TBM's shortcomings accrue greater nuance when it is required to account for the epistemic value implicit in *scientific* practice. In what follows, we present three of these challenges—all concerning whether TBM imposes licit *necessary* conditions on epistemic value—and show that ITM succeeds where its forebear fails.

6.5.1 Non-Propositional Representations

Both TBM and ITM require epistemic value's locus to be propositionally structured. By contrast, the vehicles of scientific representation are theories, models, diagrams, and so forth. None of these representations are naturally glossed as propositions. As we saw, TBM has three ways of accounting for epistemic value. We shall now argue that the basic and componential TBM strategies fail to account for scientific representations' epistemic value, and that TBM's instrumental strategy is less plausible than its ITM counterpart. Moreover, ITM's greater plausibility in this context is an instance of a broader argumentative strategy.

Begin with TBM's basic strategy, which holds that these sundry scientific representations are in fact true beliefs. As mentioned above, scientific representations are frequently non-propositional. On the leading accounts, scientific representations are models. For those of a more structuralist bent, models are set-theoretic structures (van Fraassen 2008). Such structures are neither true nor false, and, for this reason, are not propositional. However, even non-structuralist accounts of scientific representation renounce the idea that models must be propositional (Giere 2006, 64–65; Suárez 2012, 216–217; Weisberg 2013, 22).

TBM's componential strategy looks equally unpromising, for true beliefs are not components of scientific representations. Indeed, in the scientific inquiries that do result in belief, the exact opposite is true: scientific representations are the epistemically valuable components of the relevant beliefs. Since scientific models frequently lack propositional structure, one cannot simply 'believe a model'. Rather, one must believe *that* a model is, for example, accurate. However, this implies that models are contents—'components', so to speak—of their attendant beliefs. Parallel points apply to other scientific representations, such as theories and measurements.

While TBM's basic and componential strategies are nonstarters, its instrumental strategy contains a grain of truth. On this line, such representations are not goals of inquiry unto themselves. Rather, they are a means to acquiring true beliefs.[10] We agree that, insofar as scientific representations are non-propositional, they are merely a means to fulfilling inquiry's goals. However, our instrumental strategy suggests that representations are not a means to acquiring true beliefs, but a means to providing true answers to relevant questions. For instance, x-rays are a means to answering the question, "Is the patient's bone broken?". In a similar vein, statistical mechanics is a means to answering questions such as, "How do ions, atoms, and molecules vibrate within crystals?", "How do intermolecular forces affect the thermodynamics of gases?" and so on.

We can clarify ITM's instrumental strategy. Even when construed non-propositionally, scientific representations are widely thought to support "surrogative inference"; very roughly, competent and informed users can take propositions about the representational source as premises and draw valid conclusions about the target (Suárez 2002). Surrogative inferences are sound when their conclusions are true. Thus, a scientific representation, A, of a target, B, is epistemically valuable in context C just in case the sound surrogative inferences from A to B have conclusions that correctly answer every relevant question in C. In this way, ITM's instrumental strategy accommodates non-propositional representations. Moreover, Sects. 6.5.2 and 6.5.3 show that questions' relevance—and, by implication, ITM's perspectivism—is crucial in accounting for different modeling practices' epistemic value.

We think that ITM's instrumental strategy is more plausible than TBM's. Section 6.4 provides a simple explanation: precisely because ITM better characterizes inquiry's goals than TBM does, ITM's instrumental strategy will outperform its TBM counterpart (*ceteris paribus*). Specifically, TBM's instrumental strategy counsels scientists to pursue the most effective means to acquiring true beliefs, regardless of whether those true beliefs answer any relevant questions. Thus, just as *non sequiturs* were problems for TBM in Sect. 6.4, non-propositional representations that are a means to these *non sequiturs* will pose problems for TBM's instrumental strategy in this discussion. Similarly, just as true answers that were accepted or publicly implemented posed problems for TBM above, non-propositional representations that yield true answers to relevant questions but are 'merely' accepted or publicly implemented will be denied the epistemic value that they deserve. Nor are these problems peculiar to non-propositional representations: they will arise in any case where ITM's strategy is (*ceteris paribus*) the same as TBM's. Whenever this arises, we'll call it ITM's *axiological advantage*.[11]

---

[10] Of course, scientists sometimes treat these representations as goals of inquiry. Insofar as they do, such representations are proximal or subgoals of a more fundamental or overarching goal. TBM takes this more fundamental goal to be true beliefs; ITM, true answers to relevant questions.

[11] Ahlstrom-Vij & Grimm (2013) hold that accurate representations, rather than true beliefs, are of fundamental epistemic value. While their view ably handles non-propositional representations, it requires an individual mental state—the "grasp" that is characteristic of understanding. Hence, their view cannot account for public answers, and thereby does not fully overcome ITM's axiological advantage. Furthermore, we are skeptical that such non-propositional grasping is essential to understanding (Khalifa 2017, 72–79).

To summarize, science provides some of our most exemplary kinds of inquiries. Frequently, these inquiries attain their goal by hitting upon accurate representations—theories, models, and the like—that are not propositions. According to ITM, such representations are a means to answering relevant questions, and are thereby of instrumental epistemic value. By contrast, TBM does not readily account for these representations' epistemic value. Hence, only ITM accords with these aspects of scientific practice.

6.5.2   Idealizations

Critics frequently point to idealizations as evidence against TBM. In short, idealizations appear to be epistemically valuable falsehoods, and this is thought to be incompatible with TBM. Once again, TBM's basic and componential strategies are unpromising, and its instrumental strategy succumbs to ITM's axiological advantage. Begin with a shopworn example. The ideal gas law is:

$$pV_m = RT$$

Here, p, V_m, and T denote a gas' pressure, molar volume, and temperature, respectively, and R is the ideal gas constant. In some statistical-mechanical derivations of this law, particles in a gas are assumed not to interact.[12] Though false, this assumption appears to advance our understanding of gases.

Such understanding is epistemically valuable, yet TBM does not readily accommodate it. Quite clearly, the basic TBM strategy will not work, because the proposition *that particles in an ideal gas do not interact* is false and (consequently) acceptance rather than belief appears to be a more appropriate attitude toward such a proposition (Elgin 2017, Doyle et al. forthcoming). Of course, this also means that true beliefs are not components of the idealization (or the attitude thereof), so the componential strategy also fails.

TBM's instrumental strategy suggests that idealizations are a means to acquiring true beliefs.[13] For example, assuming non-interacting particles makes the ideal gas law's underlying statistical mechanics more salient. As before, this contains a grain of truth that is better articulated by ITM's instrumental strategy, which, when applied to idealizations, recapitulates ITM's axiological advantage. So, for the reasons discussed above, idealizations are better regarded as a means to answering relevant questions than populating scientists' heads with true beliefs.

---

[12] NB: Some authors claim that the ideal gas law assumes that particles do not interact. That is imprecise at best and incorrect at worst. The ideal gas law is a macroscopic law and thereby is altogether silent about whether gases are composed of particles. Rather, statistical-mechanical models from which the ideal gas law can be derived make assumptions about particles (and their interactions). Some of these models assume that particles interact; others do not. See Doyle et al. (forthcoming) for a discussion.

[13] Doyle et al. (forthcoming) argue this point.

Additionally, true-belief monists who assign mere instrumental epistemic value to idealizations must address two crucial issues that ITM ably resolves. First, are idealizations epistemically benign, i.e., of *no* epistemic value, or are they epistemically harmful, i.e., of *negative* epistemic value owing to their falsehood? Second, if idealizations are epistemically harmful, then what kind of principled "axiological book-keeping" assures true-belief monists that idealizations' positive epistemic value outweighs their negative epistemic value?[14]

Shifting from TBM to ITM fills these gaps. Begin by distinguishing epistemically benign and epistemically harmful falsehoods. Our view suggests that epistemically benign falsehoods are false answers to *irrelevant* questions. This does justice to the assumption of non-interacting particles in deriving the ideal gas law. Consider the central question in this example—why does the ideal gas law hold? As several authors note, whether particles interact is no part of the answer to this question (Doyle et al. forthcoming; Strevens 2008; Khalifa 2017; Sullivan and Khalifa 2019). Rather, the partition function—which is true—does the lion's share of the work.[15] So, in many contexts in which this why-question is relevant, an *irrelevant question* would be, "Do particles in an ideal gas interact?". Thus, the assumption of non-interacting particles is epistemically benign because the only false answers it yields are to irrelevant questions such as this one.

We can contrast this with epistemically harmful falsehoods, which have negative epistemic value because they are false answers to *relevant* questions. Continuing with our example, a false answer to the question of why the ideal gas law obtains, for instance, would be epistemically harmful. Indeed, cases in which questions about particle interactions become relevant are readily available. For instance, a slightly more sophisticated equation of state than the ideal gas law is the van der Waals equation:

$$\left(p + a / V_m^2\right)\left(V_m - b\right) = RT$$

Here, a and b represent intermolecular attraction and molecular volume, respectively. Importantly, $a \neq 0$. Thus, whereas a false answer to the question, "Do particles in this gas interact?" makes no difference to answering why the ideal gas law obtains, the same cannot be said when answering why the van der Waals equation holds. In short, the latter question erotetically implies the question about particle interactions. Thus, in the case of the van der Waals equation, the falsehood *that particles do not interact* is epistemically harmful.

---

[14] NB: Some deny that idealizations have *any* epistemic value, but are replete with *non-epistemic* benefits, such as easier calculation (e.g., Sullivan & Khalifa 2019). Such views face similar challenges, for they may need to keep axiological books on whether idealizations' positive *non-epistemic* value eclipses their negative epistemic value. Our arguments against TBM's instrumental strategy apply, with minor revision, to these positions.

[15] A staple of statistical mechanics, the partition function Z is given by a sum over all states of the system in terms of the energy E of each state: $Z = \Sigma e^{-E/kT}$.

We have seen that ITM fruitfully distinguishes between epistemically benign and epistemically harmful falsehoods. It thereby obviates any "axiological book-keeping". Since the idealization involved in the ideal gas law is epistemically benign, whatever positive value it possesses can shine through at no epistemic cost. Moreover, this once again rests on which questions are relevant. Hence, it provides another advertisement for ITM's perspectivism.

Could TBM pull off an analogous move? On such a view, epistemically benign falsehoods are not believed (but are perhaps accepted), while epistemically harmful falsehoods are believed. However, this TBM proposal looks deeply flawed. Consider two scenarios that are identical, save for the following:

False Belief: Jack believes a false answer to a relevant question, say that little demons' machinations are why the ideal gas law obtains.

False Acceptance: Jack accepts, but does not believe, the same false answer to the same relevant question.

If this TBM proposal is correct, then *False Belief* is epistemically harmful but *False Acceptance* is benign. However, this just seems wrong. By contrast, ITM delivers the more plausible verdict: both situations are epistemically harmful, as both provide false answers to relevant questions. Furthermore, because they cannot satisfactorily distinguish between epistemically benign and epistemically harmful falsehoods, true-belief monists must still balance their axiological books.

To summarize, we have argued that TBM's advocates are right to think that ide-alizations are of mere instrumental epistemic value, but are wrong to think that true beliefs ground this epistemic value. Any account that takes idealizations to be of instrumental epistemic value must distinguish between epistemically benign and epistemically harmful falsehoods. Our way of funding this distinction outdoes TBM's. Hence, idealizations are more profitably understood as an effective means to answering relevant questions—as determined by inquirers' perspectives.

6.5.3 Approximations

Idealizations vividly illustrate TBM's poor fit with one of model-based science's representational tropes. Might TBM fare better with more mundane tropes of these kinds? Specifically, idealizations are deliberate distortions, but even the most accu-rate scientific representations are approximations. As we shall now show, ITM bet-ter explains approximations' epistemic value.

In approximations, something is close to the truth, but not perfectly accurate. For example, in using the ideal gas law, scientists appear to answer questions such as:

Q1. How much will doubling pressure affect temperature?

Their answer is:

A1. Doubling pressure will double temperature.

Precisely because the ideal gas law does not countenance particle interactions, this answer is only *approximately* true. However, all such approximations are, *strictly speaking*, false. Furthermore, because A1 is false, approximations would appear to pose problems for both TBM and ITM. Indeed, for reasons analogous to those discussed with idealizations, no simple application of the basic and componential TBM/ITM strategies will work, since there simply are no truths expressed by approximations. Nevertheless, the basic ITM strategy is not doomed. By highlighting the role that background assumptions play in determining speakers' questions (and hence the range of possible answers), we can recover true answers from strictly false statements. For instance, scientists know that the ideal gas law is an approximation that only works at low pressure and high temperature. These standards of approximation are part of the implicit common knowledge operative in most scientific contexts. Thus, competent and informed audiences will interpret Q1 as shorthand for the following, more explicit question:

Q2. At *low density and high temperature,* how much will doubling pressure affect temperature *within an acceptable margin for error (ε)?*

Similarly, such audiences will interpret A1 as expressing:

A2. At *low density and high temperature,* doubling pressure will double temperature *within an acceptable margin for error (ε).*

Crucially, A2 is not merely approximately true, but strictly true. Thus, whether a putative approximation is epistemically valuable depends on the phenomena being studied and the purposes to which it is being requisitioned. Questions about the phenomena nicely capture these dimensions of approximation, and can thereby do justice to approximations' shifting fortunes regarding epistemic value.

By contrast, TBM has no such mechanism. To see this, note that both the ideal gas law and the van der Waals equation can be regarded as approximations of the state of affairs more accurately represented by the virial equation of state:

$$\frac{pV_m}{RT} = 1 + \frac{B}{V_m} + \frac{C}{V_m^2} + \frac{D}{V_m^3} + \ldots$$

This expansion is rendered arbitrarily precise by extending the equation indefinitely, with each added term being derivable from increasingly detailed and accurate assumptions about the intermolecular forces. For instance, B corresponds to interactions between pairs of molecules; C, triplets; D, quartets; etc. Every non-virial equation of state, such as the ideal gas law and the van der Waals equation, is approximately true under different boundary conditions, and is, strictly speaking, false.

TBM may attempt to accommodate these and other approximations by mimicking our strategy. Such mimicry would distinguish between: (a) the true claim that *approximately, p* and (b) the false but approximately true claim that *p.* Since it builds the approximation into the content of the proposition, so to speak, the former is true and not merely approximately so. Hence, just as ITM would allow (a) to serve as an answer, TBM would allow it to serve as a belief. In this example,

$pV_m = RT$ is false, but a nearby claim is true, and not merely approximately so. Perhaps what is believed is the truth that, *at low pressure and high temperature*, $pV_m \approx RT$ (Mizrahi 2012). This suffices as far as it goes, but, once again, ITM explains approximations' epistemic value better than TBM.

For instance, the ideal gas law (or van der Waals equation, for that matter) is simply not epistemically valuable when it comes to the purposes that other equations of state serve. Consider the stiffened equation of state,[16] which has many applications, e.g., modeling underwater nuclear explosions. Additionally, it has practical applications such as sonic shock lithrotripsy—a treatment for kidney stones and other ailments. Since all equations of state are approximations of the virial expansion, a TBM strategy that simply swapped out "=" for "≈" fails to explain why the ideal gas law is not epistemically valuable in modeling these phenomena. By contrast, ITM has no such problem: only the stiffened equation of state correctly answers questions about underwater nukes, kidney stones, and other delights.

Moreover, this case is not isolated. Dozens of equations of state exist, and figure in the modeling of explosives, seawater salinity, stars, the products of particle interactions, oilfield reservoirs, and so on. Each phenomenon is the object of a distinct line of inquiry with its own set of questions and concomitant background assumptions. Thus, ITM is superior to TBM in accounting for why different approximations are epistemically valuable in different circumstances, and precisely because of its perspectivism. Additionally, and as before, equations of state earn their keep by being answers to questions, and how they are implemented in scientists' minds is secondary—yet another manifestation of ITM's axiological advantage.

However, perhaps TBM's instrumental strategy provides a reprieve. This would mean that, e.g., accepting the ideal gas law is valuable simply because it is a means to achieving a true belief. This will also succumb to the axiological advantage, but even if we bracket that point, there is a further question: to which truths are equations of state a means? As we see it, the two most plausible options fail to redeem TBM.

The first option is that non-virial equations of state are each a means to the virial expansion; to use Elgin's (2007, 41) apt turn of phrase, they're mere "way stations" to something more accurate. However, this gets scientific practice backwards: much of scientific discovery in this area uses the virial expansion to discover new equations of state that model more specific phenomena. The ideal gas law's historical peculiarity and simplicity obscure this fact. By contrast, the stiffened equation of state is more representative: it was discovered by conjoining the virial expansion with assumptions about highly pressurized water's physical properties and then performing the appropriate derivations. Thus, the stiffened equation of state is not plausibly regarded as a means to discovering the virial expansion, since this gets the order of discovery and derivation backwards. Indeed, this suggests that the virial expansion is a means to answering questions about the stiffened equation of state (though not merely so).

---

[16] The stiffened equation of state is: $p = \rho(\gamma - 1)e - \gamma p^0$. Here, ρ is the water's density, e is the internal energy per unit mass, γ is an empirically determined constant (≈ 6.1), and p^0 is another constant, representing the attraction between water molecules.

On the second way of glossing TBM's instrumental strategy, the ideal gas law is a means to acquiring true beliefs about changes in particular gases' pressure, volume, temperature, etc. for gases at low pressure and high temperature. For this to work, the propositions about changes in a gas' properties would have to be true, and could not be mere approximations. Otherwise, they simply reignite the fires they were supposed to extinguish. However, this is implausible. If pressure *approximately* increases by a given magnitude, then one cannot infer that temperature increases *exactly* by a proportional magnitude. Of course, true-belief monists could avoid this result by hedging these claims using "≈" instead of "=", but we have already rehearsed ITM's advantages on this front.

Thus, like idealizations, ITM surpasses TBM against the problems posed by scientists' ample use of approximations. Specifically, it more precisely indicates when a particular approximation is epistemically valuable. Unlike our previous discussions, however, approximations are not of mere instrumental epistemic value; they can serve as goals of inquiry unto themselves. This is as it should be: many answers to scientific questions traffic in approximations.

6.6   Perspectivism Defended

Summarizing, inquisitive truth monism surpasses true-belief monism in better characterizing both scientific inquiry's goals and non-propositional representations, idealizations, and approximations' epistemic value. In making our case, we have let interests, social roles, and background assumptions (whether true or false) determine inquirers' perspectives. Perspectives, in turn, determine which questions are relevant to inquirers and answers to those questions are the locus of fundamental epistemic value. All of inquisitive truth monism's strengths rely on its distinction between relevant and irrelevant questions and thereby rely on its perspectivism.

Of course, with perspectives as our prime movers, some natural worries arise. Intuitively, perspectives based on epistemically bankrupt "perspectival factors" i.e., interests, social roles, and background assumptions, accrue less epistemic value than the scientific perspectives described above. Such an objection takes several forms, none of which undercut ITM.

First, such objections might saddle ITM with the commitment that an answer must cohere with a perspective in order to be correct. This is clearly mistaken, for ITM accords no epistemic value to *false* answers. Furthermore, Sect. 6.5.2 suggests that ITM ought to accord negative epistemic value—epistemic harm—to false answers to relevant questions. Importantly, we do not assume any exotic perspectival theory of truth—the T-schema does just fine. Hence, although new age claptrap, lies, and mistaken claims are put forward as answers to relevant questions from particular perspectives, those answers lack epistemic value. Similarly, we assume that x is a means to y only if x causes or raises the objective probability of y. Since ITM only concerns epistemic value, it need not regard facts about causation and objective probability as perspectival. Thus, like truth, instrumental epistemic value is not wholly at the mercy of quirky perspectives. As a result, our view accords no epistemic value to false answers or objectively unreliable methods, regardless of their centrality to misguided perspectives.

Alternatively, one may worry that some relevant questions are not worth asking because the background assumptions that support their erotetic implications are false. We see no reason why false background assumptions from which questions are (erotetically) inferred would raise such worries. For instance, Newtonian mechanics was a fruitful theory that, in conjunction with its superquestions, e.g., "How do objects move?" erotetically implied many other questions, e.g., "Why does the perihelion of Mercury precess?". The latter question's relevance to Newtonian physicists and their successors is incontrovertible. Furthermore, although such questions were only correctly answered by abandoning Newtonian mechanics in favor of relativistic mechanics, the true answer to this question is clearly epistemically valuable, just as ITM states. Thus, false background assumptions *per se* are no hurdle to relevant questions or epistemic value.

However, this raises a deeper worry. Among these assumptions are the inquirer's pressuppositions, i.e., those statements a speaker commits herself to when asking a question. In erotetic logic, presuppositions are any statement entailed by each answer to a question. For instance, if someone asks, "Who drank all the whiskey?", she presupposes that someone drank all the whiskey. This presupposition is part of the background assumptions that inform the speaker's perspective, which, in turn, determines what questions are relevant for her.

In the Newtonian example, some background assumptions but no presuppositions were false. But a more pressing objection arises when questions' presuppositions are false, for by ITM's own standards, such questions admit of no true answer. This seems to provide compelling grounds for claiming that questions with false presuppositions cannot be relevant. We disagree, for there are two kinds of answers, corresponding to whether a question has true or false presuppositions.

Thus far, we have focused on the former case, where *direct* answers are apt. Roughly put, a direct answer is a response that provides neither more nor less information than its question demands (Belnap and Steel 1976). For instance, the proposition, *that Jim only went to the movies*, directly answers the question, *Did Jim go to the mall or to the movies?* However, when questions have false presuppositions, true answers will be *corrective*; not direct. Consider, for instance, a case in which two parents are reluctant to vaccinate their children. They ask their pediatrician, how the measles, mumps, rubella (MMR) vaccine causes autism in children. The pediatrician replies, "I'm sorry, but you're mistaken. MMR vaccines don't cause autism". Here, she does not provide a direct answer to the parents' question, but instead corrects it by denying one of its presuppositions. Corrective answers to misguided questions are epistemically valuable, for we learn something true that we did not know before. Importantly, this requires no revision to ITM, as inquirers should always provide a true answer. It simply specifies that some of these answers are direct and others corrective.

Finally, one may worry that our perspectivism conflates epistemic with non-epistemic value. Throughout we have assumed that a truth's ability to answer a relevant question is of *epistemic* value. However, our perspectival factors are intuitively of non-epistemic value. Perhaps, then, a truth's ability to answer a relevant question is of *non-epistemic* value. On this view, truths that fail to answer relevant questions are epistemically valuable, and any negative valence we associate with them is merely because of their lack of non-epistemic (practical, aesthetic, etc.) value.

Notice that by assuming our perspectival factors are of non-epistemic value, this objection raises precisely the metaepistemological worries mentioned in Sect. 6.3: what theory-neutral evidence underwrites this intuition? However, even if we bracket this for the sake of argument, it does not follow that the ability to answer a relevant question is of non-epistemic value. Consider: the fact that a has property F and that a also determines b does not entail that b is F. For instance, the fact that one's parents are born in Egypt and that one's parents determine whether or not one was born does not entail that one was born in Egypt. By parity of reasoning, even if our perspectival factors are of non-epistemic value and determine whether a truth answers a relevant question, it does not follow that a truth's ability to answer relevant questions is of non-epistemic value.

Thus, all told, ITM's hearty embrace of perspectivism is defensible. It requires *true* answers to relevant questions to underwrite epistemic value, and requires no exotic perspectivism about truth, causation, or probability. Its allowance of false background assumptions, including false presuppositions, to yield relevant questions seems to be a feature rather than a bug, for such questions can be engines of good inquiry, either because they erotetically imply questions with true presuppositions (as in the case of Newtonian mechanics and Mercury's perihelion) or because they lead to corrective answers that reveal where past inquiries have gone awry. Finally, we have not covertly changed the subject from epistemic value to non-epistemic value.

6.7 Conclusion

To conclude, we have argued that inquisitive truth monism—the claim that only true answers to relevant questions are of fundamental epistemic value—outperforms the more venerable true-belief monism. By exhibiting greater fidelity to the Intuition About Inquiry, our view more readily accounts for the non-propositional and intersubjective dimensions of scientific representation, as well as idealization and approximation's epistemic value. Furthermore, since questions' relevance is a function of inquirers' interests, social roles, and background assumptions, our view entails that epistemic value is inherently perspectival.

Moreover, ITM suggests several exciting lines of development. Most obviously, we would like to argue that being a true answer to a relevant question is not just necessary but also sufficent for being epistemically valuable. Additionally, we have only compared ITM to TBM. However, favorable comparisons with epistemic value pluralists and monists of different persuasions would cement ITM's plausibility.

Similarly, our perspectivism invites comparisons with other prominent perspectivist positions in the philosophy of science (e.g., Massimi 2018).

Finally, our view raises questions about how different social arrangements might yield different allotments of epistemic value. Are people in certain social roles not entitled to ask or answer questions that would be epistemically valuable to them, given their broader interests? Might they sometimes be forced to answer questions that, while epistemically valuable to their audiences, are morally harmful to them? In this way, we see perspectivism about epistemic value as promoting a commitment to interrogate—and hopefully prevent—various kinds of epistemic injustice.

Acknowledgments We would like to thank Michela Massimi, Suilin Lavelle, Kate Nolfi, and the audience at the Edinburgh conference, *Knowledge from a Human Point of View*, for their feedback on earlier drafts.

Bibliography

Ahlstrom-Vij, K., & Grimm, S. R. (2013). Getting it right. *Philosophical Studies, 166*(2), 329–347.

Ahlstrom-Vij, K. (2013). In defense of Veritistic value monism. *Pacific Philosophical Quarterly, 94*(1), 19–40.

Belnap, N., & Steel, B. (1976). *The logic of questions and answers*. New Haven: Yale University Press.

Berker, S. (2013). Epistemic teleology and the separateness of propositions. *Philosophical Review, 122*(3), 337–393.

Bird, A. (2010). Social knowing: The social sense of 'scientific knowledge'. *Philosophical Perspectives, 24*(1), 23–56.

Buchak, L. (2014). Belief, credence, and norms. *Philosophical Studies, 169*(2), 1–27.

Cohen, J. (1992). *An essay on belief and acceptance*. Oxford: Clarendon Press.

David, M. (2001). Truth as the epistemic goal. In M. Steup (Ed.), *Knowledge, truth, and duty* (pp. 151–169). New York: Oxford University Press.

Doyle, Y., Egan, S., Graham, N., & Khalifa, K. (forthcoming). Non-factive understanding: A statement and defense. *Journal for General Philosophy of Science*. https://doi.org/10.1007/s10838-019-09469-3

Elgin, C. (2007). Understanding and the facts. *Philosophical Studies, 132*(1), 33–42.

Elgin, C. (2017). *True enough*. Cambridge: MIT Press.

Giere, R. N. (2006). *Scientific perspectivism*. Chicago: University of Chicago Press.

Goldman, A. I. (1999). *Knowledge in a social world*. Oxford: Oxford University Press.

Groenendijk, J., & Stokhof, M. (1984). *Studies on the semantics of questions and the pragmatics of answers*. Dissertation. University of Amsterdam.

Hamblin, C. L. (1958). Questions. *Australasian Journal of Philosophy, 36*(3), 159–168.

Hart, H. L. A. (1968). *Punishment and responsibility: Essays in the philosophy of law*. Oxford: Oxford University Press.

Hieronymi, P. (2008). Responsibility for believing. *Synthese, 161*(3), 357–373.

Khalifa, K. (2017). *Understanding, explanation, and scientific knowledge*. Cambridge: Cambridge University Press.

Kukla, R. (2015). Delimiting the proper scope of epistemology. *Philosophical Perspectives, 29*(1), 202–216.

Lynch, M. P. (2004). *True to life: Why truth matters*. Cambridge: MIT Press.

Magnus, P. D. (2013). What scientists know is not a function of what scientists know. *Philosophy of Science, 80*(5), 840–849.

Massimi, M. (2018). Four kinds of perspectival truth. *Philosophy and Phenomenological Research, 96*(2), 342–359.

Merton, R. K. (1942/1973). The normative structure of science. In *The sociology of science: Theoretical and empirical investigations* (pp. 267–278). Chicago: University of Chicago Press.

Millson, J. (2014). Queries and assertions in minimally discursive practice. In Rodger Kibble et al. (eds.), *Proceedings of the society for the study of artificial intelligence and the simulation of behavior, AISB'50.* Goldsmiths College, UK.

Millson, J. (2019). A cut-free sequent calculus for defeasible erotetic inferences. *Studia Logica.* https://doi.org/10.1007/s11225-018-9839-z.

Millson, J. & Khalifa, K. (in press). Explanatory obligations. *Episteme.*

Mizrahi, M. (2012). Idealizations and scientific understanding. *Philosophical Studies, 160*(2), 237–252.

Olsson, E. J. (2007). Reliabilism, stability, and the value of knowledge. *American Philosophical Quarterly, 44*(4), 343–355.

Pedersen, N. J. L. L. (2017). Pure epistemic pluralism. In A. Coliva & N. J. L. L. Pedersen (Eds.), *Epistemic pluralism* (pp. 47–92). Cham: Springer.

Pritchard, D., Turri, J. & Carter, J A. (2018). *The value of knowledge.* The Stanford Encyclopedia of Philosophy. https://plato.stanford.edu/archives/spr2018/entries/knowledge-value.

Sosa, E. (2003). The place of truth in epistemology. In L. Zagzebski & M. DePaul (Eds.), *Intellectual virtue: Perspectives from ethics and epistemology* (pp. 155–180). New York: Oxford University Press.

Strevens, M. (2008). *Depth: An account of scientific explanation.* Cambridge: Harvard University Press.

Suárez, M. (2002). An inferential conception of scientific representation. *Philosophy of Science, 71*(5), 767–779.

Suárez, M. (2012). The ample modelling mind. *Studies in History and Philosophy of Science Part A, 43*(1), 213–217.

Sullivan, E. & Khalifa, K. (2019). Idealizations and understanding: Much ado about nothing? *Australasian Journal of Philosophy.* https://doi.org/10.1080/00048402.2018.156433.

van Fraassen, B. C. (2008). *Scientific representation: Paradoxes of perspective.* Oxford: Oxford University Press.

Weisberg, M. (2013). *Simulation and similarity: Using models to understand the world.* Oxford: Oxford University Press.

Wiśniewski, A. (1995). *The posing of questions: Logical foundations of erotetic inferences.* New York: Springer.

Wiśniewski, A. (2013). *Questions, inferences, and scenarios.* London: College Publications.

Wray, K. B. (2001). Collective belief and acceptance. *Synthese, 129*(3), 319–333.

Chapter 7
Perspectivalism About Knowledge and Error

Nick Treanor

Abstract Knowledge and error have a quantitative dimension – we can know more and less, and we can be wrong to a greater or lesser extent. This fact underpins prominent approaches to epistemic normativity, which we can loosely call truth-consequentialist. These approaches face a significant challenge, however, stemming from the observation that some truths seem more epistemically valuable than others. In this paper I trace out this perspectivalist challenge, showing that although it arises from a mistaken picture of the quantitative dimension of knowledge and error, when we reconceive how that quantitative dimension should be understood we find the perspectivalist challenge has survived unscathed.

Keywords Veritism · Epistemic normativity · Truth · Perspectivalism · Similarity

7.1 Introduction

Consider an omniscient being and a blank slate. One has perfect, complete, immaculate knowledge, the other none at all. Each of us is somewhere in between. Our knowledge grows and increases, both overall and with regard to specific subject matters or domains. How much each of us knows can decrease as well, both overall and about subject matters. Moreover, interpersonal comparisons are possible: you know more now than I knew as a child, and there are topics about which you now know more than me and others about which I now know more than you. All this points to the idea that there is a quantitative dimension to knowledge. Indeed, it seems built into the very concept of knowledge that it is a quantitative notion in the sense that one can have more and less of it. It is not just that one knows or fails to know, or has knowledge or doesn't; it is that one can know more or less, one can have more or less knowledge.

We also differ from the omniscient being and the blank slate in that there is, in all of us, some admixture of error. It is not just that how much we know can increase,

N. Treanor (✉)
University of Edinburgh, Edinburgh, UK
e-mail: nick.treanor@ed.ac.uk

© The Author(s) 2020
A. Crețu, M. Massimi (eds.), *Knowledge from a Human Point of View*,
Synthese Library 416, https://doi.org/10.1007/978-3-030-27041-4_7

diminish, and in principle be compared to how much some other person knows. It is that we can go wrong to a greater or lesser extent. That is again a quantitative notion, more-ness or magnitude in the domain of error.

Finally, it is common to think not only that knowledge and error each have a quantitative dimension, but that there is an intelligible sense to the idea that there is some degree to which a person is, at a time, getting the world right. It's not just that I here, now, know a certain amount and am wrong a certain amount. It is that I have a picture of the world that can in principle be assessed overall with regard to the faithfulness with which I represent the world. How faithfully I represent the world, overall, is some function of how much I get right and how much I get wrong, with increases of knowledge or true belief increasing overall faithfulness, and increases in error decreasing overall faithfulness.

The question I want to explore in this paper is whether there is an interesting sense in which quantities of knowledge and error are perspectival. I will not do the topic justice. I will also not do justice to the interesting work on perspectivalism in science explored by Michela Massimi (2016, 2018a, b) and others and how it bears on the overall question I am here interested in. Nonetheless, I hope there are some points of contact.

This will be a story in four parts, where we visit the issue of perspectivism briefly near the beginning to foreshadow a more sustained discussion near the end. First, I spend some time talking about a picture of epistemic normativity that is, I believe, rampant in contemporary philosophy. I say 'rampant' rather than 'prevailing' or 'standard' to adumbrate the fact this picture is, in my view, a sort of disease, one that has spread quietly and unnoticed. The second part of the story is the explanation or argument concerning why I think this picture of epistemic normativity is mistaken, why the prevalence of this picture is a problem rather than testimony to its soundness. The third part of the story concerns my effort to offer something better, something that preserves the best part of the picture I want to replace while avoiding its mistakes. The fourth and final part of the story focuses on the perspectival challenge, which we shall see survives unscathed through a radical reconception of a prevailing approach to epistemic normativity. What I mean will become clear in due course, but the basic idea is that the picture of epistemic normativity that I think is wrong threatens a kind of perspectivalism about epistemic normativity, *but so too* does the picture that I argue should replace it. I will elaborate and defend both of these points. The aim will not be to conclude with perspectivalism, but to throw down a challenge to my own view and to show how the spectre of perspectivalism haunts truth consequentialist approaches to epistemic normativity. This is the case, we shall see, whether one endorses the traditional interpretation of that approach or the interpretation that I propose has to replace it.

7.2 Truth-Conduciveness and Epistemic Normativity

The picture of epistemic normativity I focus on in this paper is one that is truth-consequentialist. There are two aspects to it. The first is a claim about the structure of epistemic normativity, that it is consequentialist. Roughly, the idea is that epistemic processes (or whatever the locus of evaluation is) should be assessed by appeal to how conducive they are to the achievement or existence of some epistemic good. The second adds to this structural story the claim that the good in question is the good of truth, or more perspicaciously the good of more truth and less falsehood. To see the basic idea, think of a consequentialist view in ethics that we can call Simple Hedonism, which holds that if action A leads to more net hedons than action B, A is morally better than B. Simple Hedonism has a consequentialist structure – actions should be evaluated by appeal to the goodness or badness of the states they bring about – and it makes a particular claim about what the good and bad states are – that they are pleasure and pain. This is the same sort of two-part structure that you see within truth-consequentialist approaches to epistemic normativity. As an illustration of the approach within epistemology, consider the following remarks: "A very plausible set of [cognitive] goals are the oft-cited aims of believing the truth—as much truth as possible—and avoiding error" (Goldman 1980, p. 32); or "An intellectual virtue is a quality bound to help maximize one's surplus of truth over error" (Sosa 1985, p. 227). The Goldman quotation articulates the claim that believing the truth is good, or that what is good is believing more truth and less falsehood. The Sosa claim adds to this the idea that it is by appeal to this good that intellectual virtue should be understood. Goldman, of course, also uses this good to develop reliabilism and a veritistic approach to social epistemology. This sort of approach is very common in epistemology and I cite Goldman and Sosa largely because their views are well-known and the lines quoted compact expressions of it.[1]

What I want to focus on here in Sect. 7.2 is not reliabilism, or virtue reliabilism, or even truth-consequentialist pictures of epistemic normativity more generally. Rather, I want to focus on something that is implicit in this general approach as it has been articulated or developed throughout philosophy, a dimension of the picture that is not normative but descriptive. This is a certain conception of what truth-conduciveness is, or of what it is to believe as much truth as possible and as little error as possible, to adopt Goldman's line, or of what it is to maximise one's surplus of truth over error, to use Sosa's. On this conception, there are some number of truths and some number of falsehoods and maximising truth is a matter of increasing the number of truths believed and minimising error is a matter of decreasing (or holding at zero) the number of falsehoods believed. Note that there are two separate claims here. The first is a claim about the in-principle countability of what is true

---

[1] There are interesting differences between truth-consequentialist and truth-teleological approaches to epistemic normativity that I here elide as I don't think they are relevant to the points made in this paper.

and of what is false. The second is a claim about such countability exhausting the quantitative structure of what is true and of what is false. Note that this second claim is not otiose or redundant on the first. Some things can be counted (e.g., the pains one has suffered, the wrongs that one has committed), without it being the case that counting exhausts their measure (how much pain one has suffered, or how much wrong one has committed). The second claim therefore is that not only does what is true and what is false have a countable structure, cardinality gives the correct measure on each.

These two claims are seldom explicit, but they underpin discussions of truth consequentialism throughout epistemology.[2] In Sect. 7.3, I will criticise both aspects of this conception of the quantitative dimension of truth and error. For now, let me bracket these issues to discuss an influential line of criticism that has been directed at truth-consequentialist approaches to epistemic normativity, for doing so will foreshadow more extensive discussion later of the threat of perspectivalism. The basic objection is that the good of more truth and less falsehood can't be the ground of epistemic normativity because not all truths and falsehoods are equal—some are better or worse, epistemically better or worse, than others, in the sense that some are such that believing them makes a big difference to one's epistemic awesomeness whereas others make only a little difference, or perhaps no difference at all.

Here is how Dennis Whitcomb sums up what is known as the trivial truths objection, which has been almost universally accepted:

> It is better epistemically to know deep theoretical truths about, for example, metaphysics or physics, than it is to know trivial truths such as truths about the number of grains of sand on the nearest beach. Indeed, even if one were to know a lot of trivial truths, and thereby fulfil the epistemic value of having more rather than less knowledge, one's epistemic states would still be deficient owing to their triviality (2015, p. 313).

Or as Paul Moser put it almost 35 years ago:

> It is somewhat implausible to hold that an epistemic agent should aim just to obtain as much truth and avoid as much error as possible. For it is plausible to suppose that some truths are epistemically more important than others (1985, p. 5).

I will not discuss this objection in detail,[3] but want to note two things. First, that the objection threatens not only a truth-consequentialist account of epistemic normativity, but what we can loosely call the objectivity and interest-independence of epistemic normativity. That is, it leaves us with the claim that some truths are more epistemically *important, significant,* or *interesting* than others. These adjectives are stand-in labels, of course, not accounts, but they at least suggest a worthiness that is contingent on our particular interests and cares. To be sure, one could provide an objective, interest-independent analysis of any of these, given that they are mere labels. But they are connotative labels and suggest (and are often intended to suggest) a movement from an objective, interest-independent account of epistemic normativity to a kind of perspectivalism wherein the epistemically normative

---

[2] For an elaboration and defence of this, see Treanor (2018), especially pages 1064–1067.

[3] See Treanor (2013) and (2014) for more discussion.

depends on our situation, or history, or particular and peculiar desires and interests. The second thing I wish to note about the trivial truths objection is that it is apt only if the truth that, for example, my phone number was once 416–684-0019 is the same amount of truth as the truth that, for example, all ordinary matter in the universe is composed of species of atoms with the same number of protons in their atomic nuclei. Those who give the trivial truths objection, or who are moved by it, have simply assumed that every truth is the same amount of truth, something like one unit of truth. We can see why they think this—every truth is *a* truth. How could it not be? But the fact every truth is a truth doesn't entail that every truth is the same amount of truth. We will discuss both points in more detail below. For now, note just that as standardly construed truth-consequentialism faces a perspectivalist threat, but the challenge is defused when we notice that it arises from an objection that conflates the numerosity and measure of *true sentences* with the numerosity and measure of *what true sentences express* (to wit, the truth, or true content), and mutatis mutandis for false sentences and what false sentences express.

7.3 A Problem for How Truth-Conduciveness Is Understood

This is the part of the story where I argue that the background, prevailing picture of *more truth* and *less falsehood* is mistaken. I describe this picture as background because it has not been explicitly elaborated and defended but rather implicitly assumed. I describe it as prevailing because it is so deeply-seated and pervasive that philosophers treat it as almost analytic, as capturing the meaning of the quantitative vocabulary concerning truth rather than as expressing a substantive philosophical theory of the quantitative dimension of truth. I don't want to spend a great deal of time on this part of the story, as I have argued it at length in several papers. It is important, however, to do more than merely gesture at arguments given elsewhere. The claim I want to dislodge is so engrained and implicit in our understanding that the criticism I want to make of it would probably not be intelligible if I were to merely cite arguments given elsewhere without rehearsing them in outline.

In this section, therefore, I will first outline in more detail the picture that I think is false, and then trace out two main lines of criticism. The aim will be to convey the rough shape and intelligibility of a set of criticisms that have been given in more detail elsewhere. The point of doing this will be to set up a positive story of how the measure of knowledge and error should be understood, one that, like what it supplants, threatens to be deeply perspectival.

What is the picture I think is mistaken? There are two ways to put it, which are importantly different but for the purposes of this paper I will sometimes run together, or at least not take pains to disentangle. One way of thinking of the picture is as the claim that more truth is a matter of more truths or of more true propositions, while more falsehood is a matter of more falsehoods or of more false propositions, where the 'more' in the analyses is the more-ness of cardinality. A second way of thinking of the picture is as the claim that more true belief is a matter of more true beliefs, while more false belief is a matter of more false beliefs, where again the 'more' in the analyses is the more-ness of cardinality. How exactly these views are different turns on questions about the nature of belief that are orthogonal to the issues I focus on in this paper, so I will principally speak of more and less truth, but occasionally of more or less true belief (or of more and less knowledge and error) when it seems stylistically felicitous.

As I mentioned in Sect. 7.2, there are two quite different aspects to this picture. First, there is a claim about structure, about what is true and what is false having a countable structure or a cardinality. Second, is a claim about this countable or cardinality structure exhausting the quantitative structure of what is true and what is false. The second claim could be true only if the first is, but the converse doesn't hold. This point can be easily overlooked because it is natural to think that if there is such a thing as the number of Xs, then who has more Xs is settled by counting. That observation is itself accurate, but the mistake is to assume that who has more X (or more of X) is always given by who has more Xs. Sometimes that is true – which city has more people in it is determined by whether the number of people in one city is greater than the number of people in another city. That is because 'more people' just means a greater number of people. But in other cases it is not true. Who has committed more harm, for instance, isn't entailed by who has committed more harms. The count of harms done and the quantity of harm done come apart. If we assume for a moment that truths are in principle countable, then it may be that more truth is just the same thing as more truths (plural). But that is a substantial philosophical view, not something that follows just from the fact that, for example, five truths are a greater number of truths than three truths. In this section, I will argue against both of these claims. There is compelling reason to doubt that what is true is countable or has a cardinality (and likewise for what is false). And there is compelling reason to doubt that, even if what is true and false is countable or has a cardinality, counting or cardinality gives its measure.[4]

Why should we think that what is true and what is false isn't denumerable? The main reason is because we can find nothing good to count. Our central grasp of what is true and of what is false is via sentences of natural language, and although the set of true sentences of a natural language is in principle denumerable, let us grant,[5] its cardinality is arbitrary. This is because, to put it simply, it's possible to say the same thing in different ways. For this reason, although we can (let us grant) count true sentences, they are not the right thing to count, since different arrangements of words, or arrangements of different words, would yield a different number of sentences while saying the very same thing. To zoom in on one example, the sentence

---

[4] So far in this paper I haven't carefully distinguished countability from cardinality, although they are of course different: the real numbers are not denumerable yet have a cardinality, the cardinality of the continuum. For the most part this doesn't matter in what follows and where it does I will be more precise.

[5] The concession is needed for two reasons. First, because it is hardly clear that even sentences have the identity conditions that are required. Second, because, for example, for every real number there could be a sentence that says that number is a real number; but then there would be uncountably many sentences.

'John is a bachelor' and the sentence 'Richard is male' are each one sentence, exactly one sentence, and so in a matter of speaking are one truth each. The sentence about John says more than the sentence about Richard, however, and if each is true then the sentence about John expresses more truth than does the sentence about Richard. To put this another way, the truth that John is a bachelor is more truth than the truth that Richard is male (more truth, not more true), even though each is one truth. (For simplicity I assume the proper names in each sentence have no descriptive content, or that if they do the scale of that content is identical.)

With this particular example it is tempting to think there is a straightforward containment relation that preserves the intelligibility of more truth being a matter of more truths. A truth to the effect that S is a bachelor decomposes, one might think, into a set of truths that includes the truth that S is male. We can therefore make intelligible that "John is a bachelor" is more truth than "Richard is male" by conceiving of what is said of Richard as a proper part of what is said of John. The promise of this as a general account, however, is illusory, since it is implausible that every sentence of natural language decomposes into a concatenation of sentences that are genuinely atomic in the sense of saying exactly one thing, no more and no less. If sentences did decompose into atomic sentences so understood, then we could in principle arrive at a set of atomic sentences that has a cardinality. But they don't so we can't.

The second aspect of the prevailing picture of the quantitative dimension of knowledge and error goes beyond the claim that what is true and false is denumerable to claim that, in addition, counting gives the proper measure of it. To put this another way, the second aspect of the prevailing picture is the claim that not only does the world divide into facts, to use Wittgenstein's expression, the facts into which the world divides are all the same size. There is compelling reason to doubt this, however. There are intuitive cases where one person knows much less than another even though they know the same number of truths, however we wish to understand that. The point is most easily made if we consider not knowledge simpliciter but knowledge of a restricted domain. Consider Edinburgh, the city, and two people Charlie and Zachary who wish to know it. Suppose Charlie knows some large number of truths, N, akin to these:

Edinburgh is the capital city of Scotland.
In 2019 approximately 500,000 people live in it.
It was a medieval city formed around a castle, with a defensive wall.
In the medieval period, it was crammed with people living in high rises and deep underground so that they could live within its protective wall.
As conflict between Scotland and England abated, the city expanded northward, outside the walls, with the creation of New Town, an extensive and splendid example of Georgian town planning.
Today Edinburgh New Town is almost perfectly intact and part of a UNESCO world heritage site, along with Old Town, the medieval core.
Edinburgh was the seat of the Scottish Enlightenment, a remarkable flowering of science, philosophy, literature, and culture generally.

Meanwhile, suppose Zachary knows exactly the same number of truths about Edinburgh, but the truths are akin to these:

Genghis Khan never visited Edinburgh.
Socrates didn't visit Edinburgh.
Edinburgh would not fit on the head of a pin.
Edinburgh is not identical to the number 5.
Edinburgh is not identical to the number 6.
Edinburgh is named 'Edinburgh'.
Edinburgh is not named 'Audrey Hepburn'.

If we compare who knows more about Edinburgh, the right thing to say is that Charlie does. This is because Zachary, as we imagine him, has no clue what Edinburgh is; he just knows that, whatever it is, Genghis Khan never visited it, it wouldn't fit on the head of a pin, it is not identical to the number 5, etc. What he knows about Edinburgh is not enough to distinguish it from things that are wildly unlike Edinburgh, such as the Andromeda Galaxy or Cheese Whiz. In contrast, what Charlie knows about Edinburgh homes in on Edinburgh, it is such that although there are many things he could not distinguish from Edinburgh, anything that he could not distinguish from Edinburgh would have to be a great deal like it. The takeaway is that Charlie and Zachary know as many truths about Edinburgh as each other, but Charlie knows more about Edinburgh.

The argument just given moves from amounts of knowledge to amounts of truth. This should be unobjectionable, since what would more knowledge be if not knowledge of more truth? If one is worried about this, however, here is a similar argument that talks just of truth and not of knowledge. Think of the whole truth about Edinburgh; this includes all the truths that Charlie knows about Edinburgh, all the truths that Zachary knows, and countless more that neither does. Now take just those truths that Charlie knows about Edinburgh and compare with just those truths that Zachary knows. Which of those is more truth about Edinburgh? One might here be tempted to insist that they are the same amount of truth – some number N truths's worth of truth, as it were. But we are asking this question in an effort to discern whether the number of truths gives the measure of truth, and this response simply assumes that it does. What we need to do is step back a bit and consider each body of truth, asking which seems, intuitively, to be more truth about Edinburgh. My claim is that the body of truth that Charlie knows is more. Anything that had just the properties the Charlie-truths ascribe to Edinburgh would have to be a lot like Edinburgh; but something could have the properties the Zachary-truths ascribe to Edinburgh and still be wildly unlike Edinburgh, e.g., Cheez Whiz and the Andromeda Galaxy.

I don't think this quick sketch of each line of argument is full enough to be convincing, but I hope it gives some sense to the idea that the measure of truth is not given by counting or cardinality. It is important to note that one should be on board with this conclusion even if one only finds the first line of argument persuasive. The second line of argument assumes that truths are denumerable or has a cardinality and argues that, even then, more truth is not a matter of more truths. But first line of argument, which says there is nothing good to count, is all that's needed to undermine the standard approach to understanding truth-conduciveness within epistemology.

7.4 What, Then, Is More Truth and Less Error?

In Sect. 7.2, I introduced truth-consequentialist approaches to epistemic normativity and showed how, as truth-conduciveness is standardly understood, they threaten a kind of perspectivalism about epistemic normativity. In Sect. 7.3, I showed how this standard understanding of truth-conduciveness is mistaken. In this part of the story, I want to propose something better. It will be an approach to the quantitative dimension of knowledge and error that avoids the critical problems I argued the standard understanding faces. We will see in the next Section that although this approach radically reconceives how truth consequentialism should be understood, the perspectivalist threat survives. If anything, in fact, it grows even stronger and more pressing. But we get to that in the fullness of time; for now, let us just see how the quantitative dimension of knowledge and error should be understood.

The standard conceptualisation of truth consequentialism rests on an unstated picture of the quantitative dimension of knowledge and error that takes the object of belief to be countable and cardinality to exhaust its quantitative dimension. Both aspects of that picture are wrong, as I have argued. I now want to show how we can conceive of the quantitative dimension of knowledge and error in a way that doesn't rest on or require these problematic claims. The proposal will be schematic and unsatisfying, in that it could hardly be thought to be a fully elaborated, detailed picture. But it will be well-motivated, I believe, since it will connect the issue of the quantitative dimension of knowledge and error to related foundational issues in metaphysics and philosophy of language in an illuminating way. Moreover, to the degree that mysteries survive—and they do!—they are mysteries we face generally, mysteries in metaphysics and philosophy of language that we are already stuck with. The point of this section, therefore, will not be to offer a full or final analysis of the quantitative dimension of knowledge and error, but to show how the problem collapses into a more familiar problem that pervades philosophy. This move is what will let us avoid the problematic aspects of the standard picture, but it is also, as I will show in the next section, the very thing that threatens to deepen the perspectivalist threat.

I'll introduce and outline the positive proposal with an analogy.[6] Suppose two artificial-intelligence devices are each charged with the task of building a physical duplicate of a target object, say some particular apple. Device A goes about the task, scanning the target region to discern what is there and what it is like, and then setting out to acquire appropriate materials and assemble them in an appropriate way. Device B does the same. After some period of time, both devices come to a rest, their task complete, or as well done as they are capable of doing. Each device has produced a physical object. The object that Device A produced is a spitting image of the target apple; it is visually indistinguishable to the human eye, but moreover has the very same mass, density and density distribution, a remarkably similar chemical structure, and so on. Let us suppose, in fact, that although it is not a perfect

---

[6] For more extensive discussion, see Treanor (2019), pp. 35–38.

molecule-for-molecule duplicate, it would take a team of scientists some effort to tell the apples apart, or to speak more precisely, to discern which is the apple and which is the artefact. The object that Device B produced, in contrast, is much less similar to the target apple. It is roughly the same size and shape, but is a bit taller, not quite as wide, has a pebbly rather than smooth skin, has an interior structure that is white, moist and sweet like the apple's but with the density of foam, and so on. It could be mistaken for the target apple after a superficial inspection, but generally the similarities between it and the apple are not nearly as deep and pervasive as with the object produced by Device A.

If we think about this situation with an eye to judging which device, A or B, did a better job of duplicating the apple, there is no question that Device A did. The object that it produced was not a perfect duplicate, but it was very close. The object that Device B produced hardly competes. To be sure, that object is still a remarkable achievement. This is because if one imagines the universe of possibilia, all the things that Device B could have produced, the object that B made is much more similar to the target apple than most of that; Device B could have gotten much less right and much more wrong. It might have made an electron, or a supernova, or a doppelgänger of Joan Rivers, or any of countless other things that are wildly dissimilar to the apple. The object that it made is still, however, substantially dissimilar to the target apple compared to the object that A produced. That much is clear.

Let us consider, though, whether we can say *why* this is so. By this I don't mean what it is about the two devices, how they were designed, that led one to do a better job than the other. The question concerns why it is that the object that Device A produced is better, qua duplicate of the target apple, than the object that Device B produced. That is a question not about the devices but about the relation between the two artefacts the devices produced and the target apple.

A natural answer to this question is that Device A, in making the object it did, *got more right*, or more fully, *got more right and less wrong*. It is almost as natural to make the further claim that Device A, in making the object it did, *got more things right and fewer things wrong*. This further claim, in effect, says that the object that A produced shares a greater number of properties with the target apple than does the object that B produced, while having fewer properties not in common. That is what it would be for Device A, in making the object it did, to get more things right and fewer things wrong. This further move is appealing but it is essential to recognise that it is a mistake. As Goodman pointed out decades ago, "any two things have exactly as many properties in common as any other two" (1972, 443).[7] We should agree, therefore, that Device A, in making the object it did, got more right and less wrong, but not take the further step of claiming that it did this by getting more things right and fewer things wrong, that is, by giving the object it produced a greater number of properties the apple has and fewer properties the apple doesn't have.

---

[7] Goodman, of course, took the point to be that we should be sceptical of similarity. But the momentum in philosophy has been in the other direction, to accept similarity and affirm a non-egalitarianism about properties, with some being more natural than others. See Lewis (1986), pp. 59–69.

The answer to the question of why the object that A produced is better, qua duplicate of the apple, than the object that B produced lies already in the situation as we described it. The object that Device A produced was more similar, overall, to the target apple than was the object that Device B produced. This is not a matter of the number of properties had in common. We don't know what it is a matter of (or even whether it is a matter of anything as opposed to basic). But it is the right thing to say, for it is better to say something unclear but true than something less unclear but false. Moreover, there is something else in this analogy that should be drawn out, as it will be relevant later. It's not just that similarity doesn't consist in the number of properties. It is that some properties make more of a contribution to similarity than others. This is a consequence of there being greater and lesser similarity but not a greater and lesser number of properties in common. But we can also see the force of it intuitively. The two apples on my kitchen table are substantially similar by virtue of sharing the property of being the fruit of a McIntosh tree, but not substantially similar (or at least less substantially similar) by virtue of sharing the property of having a volume smaller than a billion light years cubed.

This analogy concerns the similarity relation that holds between objects. But it is easy to see that the same issues are at stake when we think about quantities of knowledge and error. Instead of thinking of the objects that A and B produced, think of mental representations that correspond exactly, save being mental rather than real, to those two objects. One mental representation ascribes to the apple all and only the properties that the object produced by Device A has, and a second mental representation ascribes to the apple all and only the properties that the object produced by Device B has.[8] Each of those mental representations of the target apple seem to get the apple right to some degree and wrong to some degree, or to be, as I put it early in the paper, an admixture of knowledge and error. The question that concerns us is how to assess the quantitative aspect of that. We already know, from discussion earlier in the paper, that we cannot count how many truths and how many falsehoods each mental representation consists in, both because it doesn't consist in some number of truths and falsehoods at all, and because even if it did, cardinality would not exhaust the quantitative structure. We are now in a position to see that this should not have been a surprising conclusion, despite how prevalent the counting conception of the quantitative dimension of knowledge and error is. If it is widely accepted that one object is not more similar to another by virtue of sharing a greater number of properties while differing on fewer properties, why should we have ever thought that a representation of an object gets more right and less wrong by correctly representing a greater number of the object's properties while incorrectly representing a smaller number of the object's properties? To illustrate this with an example, think again of the apples on my kitchen table. We grant that the one on the left is more similar to the one on the right by being, like it, the fruit of the McIntosh tree than it is by being, like it, such that its volume is less than a billion light years

---

[8] For simplicity I am here focusing on intrinsic properties understood as those that do not differ across duplicates. The proper understanding of 'intrinsic' is vexed, but as I appeal to it only to simplify the issue for the purposes of presentation I think it can be bracketed.

cubed. Why, therefore, should we have ever thought that knowing that it is the fruit of a McIntosh tree is to know exactly as much (exactly as much truth or true content) about it as knowing that it is smaller than a billion light years cubed? This is, I think, a remarkable instance of how parallel lines in philosophy have drifted apart.[9]

7.5   Perspectivalism Renewed

Where are we? We have seen what is wrong with the counting or cardinality approach to the quantitative dimension of knowledge and error. Just as you can't make sense of how similar an artefact is to an apple by counting the properties they share and differ on, you can't make sense of how much a mental representation gets right and how much it gets wrong by counting the number of truths and falsehoods believed. It either can't be done, at all, or it can be done but is unreliable, generating the wrong answer at least much of the time. We have also seen how the question of how to understand the quantitative dimension of knowledge and error is intimately related to questions in metaphysics about the nature of similarity. What remains to be shown is a certain vulnerability to an objection, one that is similar to one that I considered at the end of Sect. 7.2 of our story. I will close the paper by drawing out and discussing this vulnerability.

　　Recall the objection, considered earlier in the paper, that insists that since some truths are epistemically more valuable than others, epistemic normativity cannot be grounded in the good of more truth and less falsehood. The idea is supposed to be that a so-called trivial truth and a so-called significant truth are each one truth, so believing each contributes equally to how much truth a person believes. But if that is the case, and it is epistemically better to believe the significant truth than the trivial truth, something else (beyond a mere increase in the amount of truth believed) has to explain the difference in epistemic normativity. This is a poor argument, despite its widespread acceptance, because it says nothing to establish that the epistemically more valuable truths are not *more truth* than the epistemically less valuable truths; that is a hidden assumption that arises from conflating the numerosity and quantitative dimension of the vehicle (declarative sentences of natural language) with the numerosity and quantitative dimension of the content that such vehicles express. Recall, however, that the objection, were it sound, would threaten the objectivity and interest-independence of epistemic normativity. This is because it wouldn't be that what is epistemically valuable or good is more truth, or more truth and less error, where facts about what is more and less truth or error are objective,

----

[9]Goodman's insistence that similarity is not a matter of shared properties was given within a framework of thinking of properties as extensional. But notice that his scepticism becomes even more vehement when you turn toward the semantic analogy by thinking of intensional properties: "The inevitable suggestion that we must consider intensional properties seems to me especially fruitless here, for identifying and distinguishing intensional properties is a notoriously slippery matter, and the idea of measuring similarity or anything else in terms of number of intensional properties need hardly be taken seriously." (1972, 444)

impersonal and interest-independent. Instead, the objection has it that what is epistemically good is more truth that is significant or important, where the correct analysis of that seems (most plausibly) to depend on our interests and cares, or at least on contingent properties about us as knowers and cognizers. The arguments in Sects. 7.3 and 7.4 of this paper show how this objection is mistaken. We *can* make sense of why so-called significant truths are more truth than so called trivial truths. It is because they are truths about properties that make a greater contribution to overall similarity. We don't need to say that each is one truth, so therefore the same amount of truth, so therefore that the epistemic difference between them needs to be explained by something other than the contribution they make to truth-conduciveness. In other words, the arguments we have considered so far in this paper salvage the idea that truth-conduciveness alone, rather than truth-conduciveness plus significance, can ground epistemic normativity.

The problem, however, is that the positive story I have offered closes this door on perspectivalism with a draught that opens another. To articulate the worry, let us turn again to Goodman. After pointing out that any two objects share the same number of properties, he points out that one could hope in response to count not just any properties, but properties that are privileged in some way:

> More to the point would be counting not all shared properties but rather only *important* properties—or better, considering not the count but the overall importance of the shared properties. Then *a* and *b* are more alike than *c* and *d* if the cumulative importance of the properties shared by *a* and *b* is greater than that of the properties shared by *c* and *d*. But importance is a highly volatile matter, varying with every shift of context and interest, and quite incapable of supporting the fixed distinctions that philosophers so often seek to rest upon it (1972, 444, italics in original).

Here the spectre of perspectivalism enters very forcefully, even more forcefully than it entered on the standard understanding of truth-conduciveness wherein more knowledge and less error is just a matter of counting. If the right way to understand more knowledge and less error is itself by appeal to similarity, as I have proposed, then a kind of perspectivalism or interest-relativity will infect not only epistemic normativity, but even the more basic question of whether one body of knowledge is or isn't *more* knowledge than another. That is a much more thorough and threatening perspectivalism.

The quotation from Goodman illustrates that what we can call 'importance' has to enter *somewhere*. It could enter after quantities-of-truth, as it does with the objection that I have been concerned to refute. That is, one might say that every truth is one truth and that amounts of truth are just numbers of truths; but then one has to explain epistemically normative differences as a function of amount-of-truth and importance-of-truth. Or one could take 'importance' to play an ineliminable role in how amounts-of-truth should be understood, as I do. In this case, nothing is needed to ground epistemic normativity other than amounts-of-truth, but 'importance', and its attendant threat of perspectivalism, is still in the picture, just lurking in a different shadow. Moreover, this threat of perspectivalism is more haunting. In the original position, one could take epistemic normativity to ultimately depend in some way on our interests and cares, but at least, one could say with relief, how much one knows, fails to know, and is wrong about are all objective facts of the matter. On the positive story I have offered, in contrast, *even that* may be deeply perspectival. That is a much worse place to be in for those who long for the lonely indifference of interest-independent objectivity, especially concerning matters as deep and central as truth and falsehood.

My aim in this paper is principally to trace out the threat of perspectivalism rather than defeat it. This is in the spirit of knowing thy enemy, as my own sympathies are anti-perspectivalist, for better or worse. That said, I want to close by indicating why I think the positive story I offer has more potential to deliver an analysis of 'importance' that isn't interest-relative or perspectival than does the original position. I will unfortunately be able to do little more than gesture at an argument I hope to develop more fully elsewhere.

Here is the thought in outline: A truth-conduciveness approach to epistemic normativity has to let 'importance' or 'significance' into the picture somewhere. It could be that amounts of truth are independent of importance/significance and they enter later, as a link between amount of truth and epistemic value. That is the standard position, so to speak. Or it could be that amounts of truth themselves are the amounts they are because of the importance/significance of the truths involved, as I prefer. If the standard position is correct, it is difficult to see how there could be an analysis of 'importance' or 'significance' that doesn't depend on our interests and cares, since importance and significance are at the level of value, and need to be understood as modifiers of value. Now, to be sure, value might be as objective as anything can be, but it is at least very plausible that if two truths are the same amount of truth but one is more epistemically valuable than another, the value has to be explained on the mind side of the mind/world relation rather than on the world side, and this brings to the fore perspectivalist explanations. The story that I prefer is very different. 'Importance' and 'significance' refer to an inegalitarianism on the world side of the relation, specifically to a hierarchy in the domain of properties. Some properties are more natural than others, in the sense in which that term has become familiar in metaphysics, and epistemic importance or significance rests not on our private cares and position but on the impersonal structure of reality.[10] To be sure, that might ultimately be unintelligible as independent of our interests and cares (as Goodman thought). But there are two related reasons to be hopeful.

First, it seems plausible or intelligible that a world bereft of persons is bereft of value, but much harder to see how a world bereft of persons is bereft of an inegalitarianism in the domain of properties. Electrons would still exist, one wants to think, and they would be similar to each other, and properly a class, by virtue of being negatively charged. But the class of things that are negatively charged or citrus fruits, or the class of things that are smaller than the moon, would be much less natural. To be sure, this is difficult to explain or understand. But it is also much harder to reject. If this is right, the inegalitarianism of amounts-of-truth is rooted in

---

[10] See Almotahari (forthcoming) for an illuminating and sustained discussion of how the quantitative dimension of truth relates to realism about the normativity of metaphysical structure.

the incurious and perspectiveless world. Second, if 'importance' or 'significance' are located in the domain of value, as the link between amounts of truth and epistemic goodness, it seems sui generis, in the sense of standing alone and requiring an explanation of its own. It is at least plausible that there is little to appeal to other than persons and their situations. In contrast, if importance/significance is located in the domain of properties, the epistemic inegalitarianism is assimilated to an inegalitarianism that is widely thought to be required to make sense of all sorts of things (e.g., but hardly uniquely, Lewis 1983).

Bibliography

Almotahari, M. (forthcoming). *Rescuing realism.*

Goldman, A. (1980). The internalist conception of justification. *Midwest Studies in Philosophy, 5*, 27–51.

Goodman, N. (1972). Seven strictures on similarity. In *Problems and projects* (pp. 437–446). New York: Bobbs-Merrill Company.

Lewis, D. (1983). New work for a theory of universals. *Australasian Journal of Philosophy, 61*, 343–377.

Lewis, D. (1986). *On the plurality of worlds.* Oxford: Blackwell.

Massimi, M. (2016). Three tales of scientific success. *Philosophy of Science, 83*, 757–767.

Massimi, M. (2018a). Perspectival modelling. *Philosophy of Science, 85*, 335–359.

Massimi, M. (2018b). Four kinds of perspectival truth. *Philosophy and Phenomenological Research, 96*, 342–359.

Moser, P. K. (1985). *Empirical justification.* Dordrecht: D. Reidel Publishing Company.

Sosa, E. (1985). Knowledge and intellectual virtue. *The Monist, 68*(2), 226–245.

Treanor, N. (2013). The measure of knowledge. *Noûs, 47*, 577–601.

Treanor, N. (2014). Trivial truths and the aim of inquiry. *Philosophy and Phenomenological Research, 89*, 552–559.

Treanor, N. (2018). Truth and epistemic value. *European Journal of Philosophy, 26*, 1057–1068.

Treanor, N. (2019). The proper work of the intellect. *Journal of the American Philosophical Association, 5*, 22–40.

Whitcomb, D. (2015). Epistemic value. In A. Cullison (Ed.), *The Bloomsbury companion to epistemology.* London: Bloomsbury Academic.

Chapter 8
Virtue Perspectivism, Externalism, and Epistemic Circularity

J. Adam Carter

Abstract *Virtue perspectivism* is a bi-level epistemology according to which there are two grades of knowledge: animal and reflective. The exercise of reliable competences suffices to give us animal knowledge; but we can then use these same competences to gain a *second-order assuring perspective*, one through which we may appreciate those faculties *as* reliable and in doing so place our first-order (animal) knowledge in a competent second-order perspective. Virtue perspectivism has considerable theoretical power, especially when it comes to vindicating our external world knowledge against threats of scepticism and regress. Prominent critics, however, doubt whether the view ultimately hangs together without succumbing to vicious circularity. In this paper, I am going to focus on circularity-based criticisms of virtue perspectivism raised in various places by Barry Stroud, Baron Reed and Richard Fumerton, and I will argue that virtue perspectivism can ultimately withstand each of them.

Keywords Virtue perspectivism · Sosa · Epistemic circularity · Virtue epistemology

8.1 Introduction

An idea that has enjoyed influential support in epistemology holds that a belief is known only if it is suitably backed up by justified beliefs that can be adduced as premises. Externalists in epistemology deny this, and doing so brings with it a straightforward way out of the ancient Pyrrhonian Problematic—albeit, one that

J. A. Carter (✉)
Philosophy, University of Glasgow, Glasgow, UK
e-mail: adam.carter@glasgow.ac.uk

© The Author(s) 2020
A. Crețu, M. Massimi (eds.), *Knowledge from a Human Point of View*,
Synthese Library 416, https://doi.org/10.1007/978-3-030-27041-4_8

123

requires an explanation: *how*, exactly, does a given belief rise to the level of knowledge in circumstances in which it is *not* backed by justified beliefs?

A different way to put this question to the externalist is: what kind of thing—if not justified beliefs—can function as a regress stopper? While it is available to both internalists and externalists in epistemology[1] to answer this question with 'experience', the externalist should hope to be able to account for how experience is suited to play such a role while at the same time steering clear of any commitment to the Myth of the Given. Those who buy into the Myth hold that experience can play such a 'regress-stopping' role only by presenting itself to a thinker in a special way, one whereby it is directly (non-inferentially) believed (on Classical Foundationalism, with *certainty*) to be present; and then the thinker can reason from this non-inferential belief to "conclusions about the world beyond experience" (Sosa 2009, 89). Wilfrid Sellars (1956) and others[2] have registered a slew of problems for proposals that rely on such a story, and I won't attempt to rehearse them here. Instead, I want to simply highlight that if the externalist is going to *avoid* the Myth of the Given, then a different kind of story altogether is needed for how experience might stop a regress. In the simple case of perceptual beliefs, this story must tell us something about how experience could bear epistemically on our justified (and known) perceptual beliefs *without* simply 'being directly apprehended with certainty'.

A straightforward externalist stance that Ernest Sosa (2009) embraces at this juncture, drawing some inspiration from Thomas Reid, holds that "[e]xperience can bear epistemically on the justification of a foundational perceptual belief by *appropriately causing* that belief" (2009, 89). This very idea might strike some philosophers with internalist leanings as deeply wrongheaded. One way to put the worry is that this stance confuses (as Richard Rorty put it) 'causation with justification'.[3] But there is a less contentious way to state the concern: the thought that a belief might be justifiably believed or known simply by being suitably caused seems to concede *nothing* to the internalist, to any extent. What the externalist should ultimately hope to do is to avoid the Myth of the Given (as well, of course, as scepticism) while doing at least some justice to internalist intuitions. Sosa (2009) has a strategy for doing all of this, and it's one that relies indispensibly on the idea of an *endorsing perspective*:

> The best way, I contend, is to distinguish between the animal knowledge that makes only minimal concessions to internalism, and reflective knowledge, with its much more substantial internalist component constituted by its requirement of an *endorsing perspective*. This perspective requires the endorsement of one's epistemic competences and with this comes an evident threat of vicious circularity (2009, 44).

On this picture, a suitable externalist epistemology must be a *bi-level* epistemology, one on which the kind of high-grade knowledge that humans ultimately aspire to is *perspectival* knowledge (even if animal knowledge is non-perspectival). In

---

[1] It's worth noting that the strategy I'm canvassing here is not available to standard forms of coherentism.

[2] For a helpful overview of proponents and critics of the mythology of the given, see Sosa (1997a).

[3] See Rorty (1979, 152).

particular, such high-grade knowledge requires one to take an endorsing perspective on one's own competences. And *here* Sosa is right to register that the threat of vicious circularity looms.[4] Epistemic competences, for Sosa, are reliable, truth-conducive dispositions such as perception, memory, reason, etc. But, as the worry goes, how could we ever arrive at the conclusion that such competences are reliable without taking for granted *that* these faculties are reliable in the course of our own reasoning?[5]

Before considering the seriousness of this kind of problem for Sosa's virtue perspectivist epistemology, and whether the view can emerge unscathed, it is worth highlighting (beyond what Sosa himself has) what is at stake here. Sosa has argued in various places throughout his career that a *non-sceptical* epistemology will have to be an externalist epistemology; only with externalism do we have the resources to respond in an adequate way to sceptical arguments from dreaming and radical deceptions.[6] That said, a plausible externalist epistemology will need to do (at least some) justice to internalist intuitions while avoiding the Myth of the Given, and *here* is where it is said that externalism must take the form of virtue perspecitivism. Now, *if* virtue perspectivism succumbs to vicious circularity, then this will lead to one of two paths: scepticism (which externalism is meant to help us avoid) or relativism, where the latter draws attention to the fact that viciously circular justification is a status that can be attained in principle through *any* kind of endorsing perspective, including those that are in conflict with one another.[7] It looks, then, like quite a lot may depend on whether virtue perspectivism succumbs to vicious circularity. In this paper, I am going to focus on circularity-based criticisms of virtue perspectivism raised in various places by Barry Stroud (2004), Baron Reed (2012), and Richard Fumerton (2004), and I will argue that virtue perspectivism can withstand each challenge.

8.2   Virtue Perspectivism and Circularity

On Sosa's virtue perspectivism, there are two kinds of knowledge: animal and reflective.

> [...] animal knowledge does not require that the knower have an epistemic perspective on his belief, a perspective from which he endorses the source of that belief, from which he can see that source as reliably truth-conducive. Reflective knowledge does by contrast require such a perspective (2009, 135).

---

[4] See Greco (2011, 105–6) for discussion.

[5] Sosa notes an historical predecessor to the problem, stated this way, in Thomas Reid's epistemology. See Sosa (2009, 62).

[6] See, for example, Sosa (1999, 2007, 2017).

[7] For discussion, see Carter (2016 Ch. 3).

What animal knowledge positively requires is just that one's beliefs are reliably truth-conductive, in that their correctness must manifest a reliable disposition of the believer. In the perceptual case, the knower's perceptual beliefs must be prompted truth-reliably by perceptual experiences. And in this way, one can come to have basic perceptual knowledge *without* relying on the kind of 'givenness' of classical foundationalism.[8] Furthermore, in securing such knowledge one needn't reason inductively from beliefs about the qualitative character of perceptual experiences.[9]

That said, let's now consider the intellectual gain that, on Sosa's view, a thinker is supposed to have secured when one transitions in the perceptual case (via satisfying the perspectival condition) from animal to reflective knowledge. A full description of this story will surely include the following admission: that in transitioning from the animal to the reflective, one would be arriving at a positive view of one's faculties that relied on, and so implicitly trusted, the use of those very faculties.

But isn't this *circular*? Or, what is more relevant: is what is described here a kind of *vicious* circularity? At this point, there is an interesting kind of *reductio ad absurdum* that Sosa first canvasses himself, and which has since become a point of focus in Sosa's dispute with Stroud. The thought is brought out when we compare an ordinary perceiver with "a crystal ball gazer who thinks that what he can see in the ball enables him to tell about matters beyond" (2009, 135). If we suppose the crystal ball indicated to the gazer a favourable view of the epistemic credentials of the crystal ball, we would not think much of the epistemic status of that favourable perspective (or what it has to offer). But—and this is the worry Sosa initially envisages for virtue perspectivism—isn't the ordinary perceiver in just the analogously same situation as the crystal ball gazer who comes to trust crystal ball gazing in that each relied on, and so implicitly trusted, the use of their respective faculties (or ways of forming beliefs) in coming to have a positive view of those faculties?

Sosa grants that the kind of epistemic circularity that is implied by the move from animal to reflective knowledge would be *viciously* circular if it made the gazer equally as justified as the ordinary perceiver. But he denies that it does, and so insists it is not. Here some care is needed because the reasoning Sosa offers for *why* the two are not on a par is not ultimately convincing to Stroud. What Sosa is willing to grant is the following: what the gazer (all going well) can attain is a kind of *internal coherence* that is every bit as equal to the internal coherence that the ordinary perceiver attains. Despite this concession, he maintains that:

> There are faculties other than reason whose apt functioning is also crucial to the subject's epistemic welfare. In light of that result, why not distinguish between the gazers and the perceivers in that, although both reason properly and attain thereby coherence and justification, only the perceivers are more fully epistemically competent and attain knowledge? On this view, the crystal-gazers differ from the perceivers in that gazing is not reliable while perceiving is. So the theory of knowledge of the perceivers is right, that of the gazers wrong. Moreover, the perceivers can know their theory to be right when they know it in large part through perception, since their theory is right and perception can thus serve as a source of

---

[8] For discussion, see Hasan and Fumerton (2017) and Carter and Littlejohn (Forthcoming Ch. 1.)
[9] This is the strategy defended by Moore (1959), and criticised in the opening sections of Sosa (2009, Ch. 9).

knowledge. The gazers are by hypothesis in a very different position. Gazing, being unreliable, cannot serve as a source of knowledge. So the perceivers have a good source or basis for their knowledge, but the gazers, lacking any such source or basis, lack knowledge (2009, 200–201).

If the foregoing is right, then Sosa can successfully sidestep the *reductio* he canvasses. For it's *not* true that the kind of circularity implied by one's move from animal to reflective knowledge would leave one epistemically on an equal footing with the crystal ball gazer. Even more, as he notes, the idea that one would need to use (and in doing so take for granted the reliability of) a given faculty at some point to come to have an adequate view of that faculty's epistemic status is implied by the very possibility of having 'an adequate theory of our knowledge and its general sources' (Sosa 2009, 196). As Sosa submits, no one could provide adequate support of any of our sources of knowing, including perception, memory, deduction, abduction, and testimony, without employing those faculties (see Sosa *ibid.*, 201.)

Sosa's response to the question of vicious circularity can accordingly be summarised as follows: the kind of circularity his reflective knower succumbs to in the case of perception is vicious only if either (i) an adequate theory of our knowledge and its general sources is impossible, or (ii) the perceiver is on an equal epistemic footing as the gazer. We have no good reason yet to think (i) is true and (ii) is disputed in the passage quoted above. And so the circularity is not vicious, but benign.

Sosa's reasoning, and in particular his way of distinguishing the epistemic plights of the perceiver as opposed to the gazer, will not seem very amenable to an internalist. Before considering Stroud's assessment of this situation, it's worth pointing out one aspect of the dialectic here that might be easily overlooked: while Sosa's distinguishing the perceiver and the gazer looks like hardened externalism, the way Sosa characterises the epistemic position of the reflective knower is not itself unconcessionary to the internalist. The reason that this point can be easily elided is that the thought experiment simply *assumes* that the gazer enjoys the same broad coherence in her beliefs as the perceiver. On Sosa's view—and this is the nod to the internalist—this broad coherence *adds* to the epistemic value of the perceiver's belief, though the value it adds is greater than the value that the gazer's coherence adds to the gazer's belief. And this is because such broad coherence (in the case of the perceiver) is truth conducive, and yields integrated understanding (even if it would *not* for the gazer, or for the perceiver were she situated in a demon world).[10] I mention this point here simply to register that *despite* the hardened externalist feel of Sosa's response to the perceiver/gazer reductio, the virtue perspectivist view advanced countenances the epistemic value of coherence (in the right circumstances) in a way that sets it apart from genuinely hardened externalist views (such as Goldman's (1999) process reliabilism or Armstrong's (1973) causal account) that make no such internalist concessions at all.

---

[10] For further discussion on this point, see Sosa (1997b, 422).

8.3   Stroud on Sosa and Circularity

According to Barry Stroud (2004), Sosa's virtue perspectivism on closer inspection
lacks the resources to suitably account for the difference between the epistemic
plights of an ordinary perceiver and a crystal ball gazer. This is quite a charge. Note
that the perceiver/gazer analogy was initially raised by Sosa himself in the context
of discussing epistemic circularity. To be clear, the worry Sosa initially envisaged
for his view (which the preceding section was a response to) went as follows: the
thinker who transitions from animal to reflective knowledge (on Sosa's view) looks
to be (prima facie) in an epistemically analogous position as a crystal ball gazer
(who relies on the ball to positively endorse its usage) in that each relied on, and so
implicitly trusted, the use of their respective faculties (or ways of forming beliefs)
in coming to have a positive view of those same faculties.

But Stroud seems to be registering a sense in which, on Sosa's view of knowledge, the ordinary perceiver turns out to be epistemically impoverished—in an
analogous way as the gazer is—*regardless* of whether either approach succumbs to
vicious circularity, per se. If this charge can be made to stick, then it's obviously a
real problem for virtue perspectivism. Let's now look at the charge more carefully.
It is levelled at Sosa against the background of a specific desideratum on an account
of knowledge in mind. This desideratum that holds that a philosophically satisfying
account of our knowledge of the external world must not *merely* be true, but it also
must be something that we know to be true. Sosa himself does not dispute the desideratum. But he, unlike Stroud, thinks that virtue perspectivism can respect it. So
why, then, does Stroud think Sosa's view *cannot* respect it? It will be helpful to
consider two key passages from Stroud:

> The question is whether holding such a theory leaves anyone in a position to gain a satisfac-
> tory understanding of knowledge of the world, even if he fulfills the conditions Sosa's the-
> ory says are sufficient for knowledge. Could someone in such a position come to recognize
> himself as knowing, and not merely confidently believing, perhaps even truly, that sense
> perception is a way of getting knowledge of the world and crystal ball gazing is not? I think
> that, on the understanding of perception that appears to be involved in Sosa's question […],
> the answer is 'No'. On that view, what we are aware of in perception is restricted to features
> of our perceptual experiences. The external facts we know as a result of those experiences
> are nothing we ever perceive to be so. What we get in sense perception therefore bears the
> same relation to the world we think we know by that means as what is seen in crystal ball
> gazing bears to the world the gazers think it gives them knowledge of (2004, 171–2).

In response to the above, Sosa briefly makes explicit that we do believe that our
perceptual experiences are reliably connected with what we think we know on their
basis—a point he registers before engaging with the following key piece of Stroud's
reasoning:

> But anyone who thinks that all it takes to have satisfactory understanding of perceptual
> knowledge is to conclude by modus ponens that we know by perception that there are exter-
> nal things would have to concede that the crystal ball gazers have a satisfactory understand-
> ing of crystal ball gazing knowledge. They could draw the corresponding conclusion
> equally confidently from what they believe about themselves (2004, 172).

The dialectic at this stage between Sosa and Stroud becomes especially thorny. Whereas Stroud is insisting that Sosa's account of knowledge is no more philosophically satisfying than the corresponding account of knowledge that might be advanced by a crystal ball gazer, Sosa for his part, finds Stroud's reasoning "hard to follow" (2009, 207). As Sosa sees things, Stroud was purporting to have granted for the sake of argument Sosa's view that perceptual knowledge *is* a matter of perceptual beliefs being prompted truth-reliably by perceptual experiences. That said, Sosa questions whether Stroud can *coherently* suppose that "if we conclude by modus ponens that we know about the world around us through perception, given that our perceptual faculties *are* reliable, then we are in the predicament he alleges?" (ibid. emphasis added). Specifically, Sosa thinks such a supposition is refuted by:

> [...] a crucial difference that Stroud and I both recognize: namely, that we know our perceptual faculties to be reliable whereas the gazers believe but do not know their gazing to be reliable. So, how can we be in an equally good epistemic position to understand how we know, if we do know but they do not know about the reliability of the faculties involved? (2009, 207)[11]

It's hard not to think that Sosa and Stroud are in some way talking past one another.

8.4 The Dialectic Between Sosa and Stroud Revisited

I want to now suggest what I think the crux of the impasse is and how, once suitably appreciated, the situation should not be especially problematic for Sosa's bi-level picture. If my interpretation of this debate is correct, we can trace the misunderstanding between Sosa and Stroud, which was reflected in the previous section, to the fact that they are thinking in very different ways about the relationship between indirect realism and externalism in the epistemology of perception. In the paper that Stroud principally takes issue with—viz., "Reflective Knowledge in the Best Circles" (reprinted in Sosa 2009, 178–210)—Sosa opens with an attack of the kind of indirect realist strategy in the epistemology of perception that he attributes to G.E. Moore, and which aims to vindicate external world perceptual knowledge as based on inference from information about our experiences. Sosa's argument against this kind of strategy takes the form of a dilemma: the relevant inference can't be deductive because "experience prompts but doesn't entail the truth of its corresponding beliefs" (ibid, 179). But an abductive strategy does no better. This is because on such a strategy we (i) "restrict ourselves to data about qualitative character of our own sensory experience" (ibid, 179) (ii) "view belief in a commonsensical external world as a theory best postulated to explain course of our experience" (ibid, 179). The problem is that if we really do restrict ourselves to just data about the qualitative character of our experiences *and nothing else* (without presupposing the external reality to be inferred), then it's unclear how the external world

---

[11] Reed (2012, 284) also discusses this particular passage in the exchange and takes Sosa's response to be perfectly consistent from within his own view.

hypothesis will beat competitor explanations (as it must do for such an abductive strategy to succeed). It is against the background of rejecting on the basis of the above dilemma the indirect realist approach that Sosa presents his own externalist strategy for vindicating external world perceptual knowledge as a favourable alternative, one on which there is claimed to be a reliable sensory basis for understanding how our perceptual beliefs can constitute knowledge.

Now—and here is where I think the crux of the dispute between Sosa and Stroud lies—notice that Sosa's own position does have in common one thing with the indirect realist approach he criticises and which he maintains falls prey to the dilemma he raises. And that is that on Sosa's view, no less than on an indirect realist view, there's a sense in which deliverances of perception are limited in a certain very general way. Let P be the proposition that *the wall is red* and let 'E' be what is available for conscious inspection whenever one has the experience of seeing a red wall. Both the indirect realist *as well* as Sosa allow that the "deliverances of perception even at its best are limited to the character of one's perceptual experiences" (Stroud 2004, 172); and so both in this respect accept that when we know a proposition like P, the deliverances of perception are limited to E.

Sosa, as we've seen, takes this commitment to be problematic *for the indirect realist* specifically because the indirect realist's *strategy* for vindicating external world knowledge (e.g., of a proposition like P) involves reasoning from E. Sosa's strategy for vindicating external world knowledge does *not* invoke any such reasoning from E. However, Stroud insists that, even so, Sosa's strategy

> *still leaves us with something that is epistemically prior to any knowledge of an independent world*. If there are no reliable connections between the perceptual experiences we receive and the world we believe in as a result of them, we know nothing of the wider world even though we know what experiences we are having (2004, 172).

I noted previously that I think the impasse between Sosa and Stroud is due principally to the fact that they are thinking in very different ways about the relationship between indirect realism and externalism in the epistemology of perception. We're now in a position, I think, to see why this is.

Just consider that, according to Stroud, it's a commitment that Sosa shares with the indirect realist that is Sosa's undoing—a commitment to accepting that what perception furnishes us with is (as Stroud puts it) something that's "epistemically prior to knowledge of an independent world" (2004, 172). Now we get to the central point. For an *externalist* like Sosa, there is no meaningful sense in which whatever perception furnishes us with is *epistemically prior* to knowledge of an external world, even if it is obviously in some way temporally or metaphysically prior.

To appreciate this point, it's helpful to consider how Sosa has put things in more recent work, where he emphasises that when one comes to possess animal knowledge, as one does in the case of simple perception, one *exercises a competence*. Crucially for Sosa, though, it is exactly the manifestation of that competence in the correctness of a belief that "thereby *constitutes* a bit of knowledge" (2017, 141). One's experiences might seem a certain way while one is exercising that competence, but it's the exercising of the competence, not the seeming, that explains the knowledge:

> what happens when we manage to open the lid and look inside. Now we may immediately know the answer to our question, with a perceptual belief—say, that there is a necklace in the box—which manifests certain cognitive competences for gaining visual experience and belief. Perhaps this complex, *knowledge-constitutive competence* first leads to things seeming perceptually a certain way, and eventually to the belief that things are indeed that way, absent contrary indications. A belief manifesting such a competence and, crucially, one whose correctness manifests such a competence, does constitute knowledge, at a minimum animal knowledge, perhaps even full-fledged knowledge (including a reflective component) (2017, 141, my italics).[12]

If perceptual seemings really were epistemically prior to knowledge, on Sosa's view, the manifestation of a competence in a correct belief could hardly constitute knowledge. And this is so even though the view that the manifestation of a competence in a correct belief constitutes knowledge is compatible with the view that we perceive at best only the character of our perceptual experiences, where the latter is irrelevant to the *epistemological* story Sosa's externalism offers for how one comes to know. And the same is the case for Sosa's story about how one knows that one knows, viz., when one attains reflective knowledge. Though reflective knowledge involves the exercise of a different kind of competence, it is *constituted* by the manifestation of (reflective) competence in a correct belief that one's first-order belief would be apt.

With this diagnosis in hand, we can now better see why Sosa does not belabour discussion of *whether* on his externalism we perceive at best only the character of our perceptual experiences. It's because this is a fact that is epistemically not significant on the kind of externalism he embraces. Now, as Stroud sees it, *attributing* to Sosa the view that on his externalism we perceive at best only the character of our perceptual experiences "seemed necessary to make sense of him as trying to answer the kind of question his 'externalist' theory is meant to answer" (2004, 172–3). My assessment is that *this* is because Stroud is reading Sosa as *needing* the character of perceptual experiences to be doing a certain bit of epistemic work in Sosa's epistemology that Sosa (in short) does not view as work that needs done.

In sum, I've suggested in this section that a kind of circularity worry that Sosa envisaged might be raised against his view has been taken by Stroud and converted into what initially looked like an even more serious epistemological defect with virtue perspectivism. Closer inspection, I think, reveals that Sosa has a satisfactory response to this worry to make from within his own externalist epistemology. While the adequacy of Sosa's response has been a source of dispute between Sosa and Stroud, this dispute itself rested on some miscommunication, where I've suggested that each has taken the other to have in mind a different conception between the (put broadly) the relationship between indirect realism and externalism in the epistemology of perception.

---

[12] See, also Sosa's discussion of metaphysical analyses in Sosa (2017, Ch. 4).

8.5 Reed on Rationalist Perspectivism and Virtue Perspectivism

According to Baron Reed (2012), Sosa's response to Stroud is stronger than it needs to be, and appreciating *why* this is so reveals a hitherto unnoticed weak spot in virtue perspectivism, one that comes to the fore when different perspectives come into contact with one another. In such circumstances, Reed thinks, there are important limits to how far virtue perspectivism can go toward giving us what Reed calls a 'resting place' for our intellectual pursuits. Or, put another way, virtue perspectivism "does give us an answer, but it is to a question that we will continue to feel the need to ask" (2012, 286). In this respect, Reed hints to an *aspect* of Stroud's objection to Sosa that is getting at something right, even if strictly speaking, Sosa is within his rights (as Reed thinks he is) to press back against Stroud's claim that Sosa's externalist can't know *that* he knows his faculties are reliable when they are.

If Reed is right, then this has important implications for virtue perspectivism as an anti-sceptical strategy. As Sosa sees it, a boon to his two-tiered account of knowledge is that it enables him to respond to the sceptic in a way that is broadly analogous, structurally, to the way Descartes thinks we should respond to the sceptic. Descartes, like Sosa, adverts to a two-tiered, perspectival structure (for Descartes, the lower level is *cognitio* and the higher-level *scientia*), where the ascent from lower to higher knowledge marks an intellectually valuable transition attained through reflection on one's epistemic position.

	Rationalist Perspectivism	*Virtue Perspectivism*
1	Begin with foundational knowledge, via rational intuition—Viz., **cognitio**; does not require knowing first that rational intuition is reliable.	Begin with foundational knowledge, via perception—Viz., **animal knowledge**; does not require knowing first that perception is reliable.
2	Acquire *more* knowledge of the same kind, directed toward one's epistemic position.	Acquire *more* knowledge of the same kind, directed toward one's epistemic position.
3	**Cognitio** now becomes something more valuable—Viz., **scientia**—By being the object of a higher order perspective (one that features broad coherence).	**Animal knowledge** now becomes more valuable—Viz., **reflective knowledge**—by being the object of a higher order perspective (one that features broad coherence).

Descartes' rationalist perspectivism, as an anti-sceptical project, is one that, *if it succeeds* (which of course is a big 'if') would secure a kind of (as Reed puts it) 'intellectual stability' that a thinker aspires to in the face of sceptical doubts. Now, as Sosa sees it, "In structure virtue perspectivism is [...] Cartesian, though in content it is not" (2009, p. 194). If Reed is right, though, this structural parallel only goes so far, and this is because crucial to Descartes' rationalist perspectivism as an anti-sceptical strategy is its *rationalist content*, in virtue of which a thinker not only can know one's faculties (for Descartes, the operative faculty being clear and distinct perception) are reliable, but attains a perspective from which she can see this with certainty, beyond doubt.

8.6 Reply to Reed

The starting place for Reed's critique of Sosa involves drawing attention to the fact that we should classify virtue perspectivism as a *fallibilist* rather than as an *infallibilist* theory of knowledge. It's fallibilist because according to virtue perspectivism, a subject can know a proposition (e.g., that the wall is red) on the basis of some justificatory source (e.g., reliable perception) even though the subject *could have had* the very same justificatory source and yet fail to know that proposition (e.g., had a jokester tinkered with the lights so that the subject was looking at a white wall bathed in red light, rather than a red wall).[13]

A key observation of Reed's is that, on a fallibilist theory, knowledge does not suffice to put to rest *doubts* we might have, in the sense that it is possible to know something (fallibly) while maintaining the epistemic possibility of the denial of what we (fallibly) know. With this point in hand, Reed observes that there is (and has been) in fact a surprising level of disagreement amongst epistemologists about what, exactly, our basic intellectual faculties *really are*. For example:

> Wilfrid Sellars includes introspection, perception, and memory. But surely that list is incomplete. Sosa would presumably add testimony and reason. Others—e.g., John Locke and Alvin Plantinga—would add a faculty of divine revelation. Yet others—the Logical Positivists, say—might want to subtract both revelation and reason. And Thomas Reid thought we have a faculty of 'common sense,' which gives us knowledge of his preferred philosophical principles. Which of these conflicting views is correct? (2012, 285)

Reed is of course right that not all of these views can be correct; in fact, at most one could be. Now—the point concerning doubt is this: if one (engaged in such a debate with others about which intellectual faculties are the basic ones and which are not) is such that her view *happens to be the correct one*, then Sosa's view permits one to claim that one knows it is correct. And even so, she may legitimately *wonder* whether it is. Such knowledge, Reed thinks, fails to provide the "healthy sort of stability that Descartes was seeking" (2012, 286), and which one would attain only if one could see with certainty that one's faculties gave one knowledge.

And so Reed's position is that even when one makes the ascent from animal to reflective knowledge, one may still have legitimate doubts and thus a lack of intellectual stability that one would not have if making the analogous ascent from *cognitio* to *scientia* within the rationalist perspectivism model. Accordingly, then, there is a sense in which virtue perspectivism's shared structure with rationalist perspectivism affords it with what are ultimately illusory anti-sceptical epistemic credentials. In response to Reed's line of critique, I'd like to make one observation and then raise two criticisms. The observation concerns the shared spirit of Reed's critique with

---

[13] This is Reed's own preferred formulation of fallibilism. See also Reed (2002).

Stroud's. Both locate what they take to be an inadequacy (at least, by the lights of broadly internalist thinking) with the epistemic position of the reflective knower insofar as she is said to know her own faculties are reliable. Stroud takes the inadequacy he locates to call in to doubt whether the reflective knower has genuine *knowledge* that her faculties are reliable. Reed, by contrast, is not contesting this point but rather calling into doubt whether a theory of knowledge that *permits* one to count as knowing her faculties are reliable while being in the position of Sosa's reflective knower is a theory that goes far enough toward meeting our intellectual needs. Given these differences, any response to Reed on behalf of the virtue perspectivist will have to take a very different shape, given that the critique itself targets a different kind of desiderata on a theory of knowledge. Having addressed this point, I'd like to now suggest how I think the virtue perspectivist might be in better shape than Reed has led us to believe.

Firstly, let us grant for the sake of argument Reed's claims about the importance of quelling doubt as a key desideratum within the project of giving a theory of knowledge. Accordingly, let's take for granted that if a subject is in the best epistemic position that a given theory of knowledge licenses, and yet legitimate doubt still persists, then the theory of knowledge has failed this desideratum. Now, even on such an assumption, Reed admits that not all disagreements Sosa's reflective knower might find herself in, about what our basic faculties are, would be likely to incite any such doubt. As he puts it:

> If the perceiver found herself only in this one disagreement, with the crystal ball gazer, she perhaps wouldn't be too badly off. Crystal ball use doesn't really have much to recommend it (2012, 285).

And it is in the context of *this* point that Reed draws our attention to more serious disputes in epistemology about what the basic faculties are, those that he thinks can legitimately leave Sosa's reflective knower with doubts. This move, though, seems to be a double-edged sword, given that it invites the counterreply that ordinary perception is common to *all* of the lists of basic faculties embraced by, e.g., Sellars, Sosa, Locke, Plantinga, Reid, and the Logical Positivists. It's accordingly not clear that the ordinary perceiver who attains reflective perceptual knowledge is going to be prompted to doubt in the face of disagreement.

A second point worth noting in response to Reed's critique of virtue perspectivism concerns his view of the importance of quelling doubt within the project of giving a theory of knowledge. On one way of reading things, this desideratum should be interpreted as a kind of necessary requirement such that it will be failed on Sosa's theory if the following situation is a metaphysical possibility: a thinker both (i) attains highest-grade knowledge; and (ii) doubts to any degree the reliability of her intellectual faculties. If this is indeed the way to think about the requirement on an account of knowledge, then virtue perspectivism fails it. But then, so arguably does rationalist perspectivism!

Here is the idea. Cartesian *scientia* can persist in two modes. In the 'active mode' one is *engaging* clear and distinct perception in taking the perspective one does on one's intellectual faculties. Such engaged clear and distinct perception is plausibly incompatible with doubt in a way that reflective knowledge is not. But what happens to *scientia* when one's mind is not so engaged? As Reed puts it, on the Cartesian picture:

> The certainty he possesses while he is entertaining his clear and distinct perceptions remains even when he is no longer entertaining them but merely remembering that they were clearly and distinctly perceived (2012, 280).

Of course, if *scientia* were available only in the active mode, then the consequence would be a radical kind of 'epistemic presentism'[14] that cedes all to the sceptic except during rare moments of engaged and self-directed clear and distinct perception. And so it's natural that a plausible articulation of rationalist perspectivism will allow *scientia* to be sustained outwith the active mode, and so through the memory of specific clear and distinct perceptions. But whereas *scientia* in the active mode is plausibly incompatible with doubt, it seems that *scientia* when sustained merely though the memory of clear and distinct perceptions is *not*. At least, it is plausibly *metaphysically possible* that *scientia* sustained through memory of the relevant kind of clear and distinct perceptions be compresent with some degree of doubt.

What this all means, is that if the relevant kind of quelling of doubt Reed takes to be a desideratum on the project of giving a satisfactory theory of knowledge is failed on Sosa's theory. This is because his theory allows for the metaphysical possibility that a thinker both attains highest-grade knowledge and doubts to any degree the reliability of her (relevant) intellectual faculties, then the same charge applies to rationalist perspectivism, and so there would be no basis for favouring the latter to the former.

A natural response to the above point would be for Reed to articulate the general doubt-related desideratum on an account of knowledge he is appealing to differently. Perhaps, rather than to say that an account of knowledge fails the doubt-related desideratum outright if it is a metaphysical possibility, on the account of knowledge, that a thinker both attains highest-grade knowledge and doubts to any degree the reliability of her intellectual faculties,[15] we might instead opt for something different. Perhaps better is the following: that for two accounts of knowledge, A and B, ceteris paribus, A is to be preferred to B if doubt is compresent with high-grade knowledge on A to a lesser extent than on B. With this kind of requirement, it might then be claimed that Sosa's reflective knower will more often find herself in a position of doubt than will someone with Cartesian *scientia*. And therefore, as this line of thought goes, Sosa's theory of knowledge lacks the resources that rationalist perspectivism does to quell sceptical doubts—and so (to conclude the argument) virtue perspectivism does not inherit the anti-sceptical import of rationalist perspectivism despite sharing its perspectival structure.

But retreating to this line is, I think, problematic for two reasons. For one thing, it's not obvious that the reflective knower is comparatively *more* inclined to doubt than is one with *scientia* sustained through memory of past clear and distinct

---

[14] See Palermos (2018).

[15] This is something we've seen rationalist perspectivism fails just as virtue perspectivism does.

perception. For another—and this is a point I don't think Reed has really addressed—it's unclear just what the reflective knower who is *lacking* any such doubts is lacking, epistemically, in comparison with the thinker with *scientia*. And given that the circumstances of disagreement Reed points to about basic faculties are not contexts where perception itself has been or is inclined to be seriously called into doubt, it would seem as though perceptual knowledge that rises to the reflective level will *de facto* be knowledge that lacks any such doubts.

For these reasons, I think that even if we grant Reed that quelling doubts is an important aspect of an epistemological project, it's not clear on closer inspection that virtue perspectivism is disadvantaged in comparison with rationalist perspectivism, or for that matter that virtue perspectivism falls short of giving us anything we should rightly expect an account of knowledge to provide.

8.7 Fumerton on Virtue Perspectivism and Coherence

The suggestion that a philosophically satisfying account of our knowledge of the external world should at least do *some* justice to internalist intuitions is one Sosa makes explicitly. And it's evident that this concession is one he views himself as making not at the animal level, but at the second-order, reflective level. But, how exactly are we best to understand this concession?

As Sosa tells us, 'broad coherence' is a feature of the second-order perspective characteristic of a reflective knower. But the coherence at the second-order is not 'untethered' coherence, but coherence that arises from a suitable provision of first-order animal knowledge. But animal knowledge *itself* is knowledge the attainment of which is accounted for by Sosa on externalist lines that are not concessionary to the internalist.

The matter of how to understand the sense in which Sosa's virtue perspectivism succeeds in doing justice to internalist intuitions seems to turn on how to think about the second-order perspective, with careful attention to how it incorporates the products of the first-order perspective.

To that end, let's take as a starting point two concise statements Sosa offers for how the first- and second-order perspectives interact. In a very recent statement, Sosa has said of his bi-level picture that it

> [...] allows the use of our basic foundational faculties in attaining a second-order assuring perspective. So we can use the animal knowledge that we attain through the exercise of such faculties; we can use such animal knowledge in the (proper, coherence-aimed) elaboration of the endorsing perspective. This endorsing perspective would be a proper awareness of our competences through whose exercise we can gain our first-order knowledge (2017, 45–46).

And previously, in *Reflective Knowledge* (1997b) he wrote

> [...] reflective knowledge, while building on animal knowledge, goes beyond it precisely through integration in a more coherent framework. This is it achieved via an epistemic

perspective within which the object-level animal beliefs may be seen as reliably based, and thus transmuted into reflective knowledge (2009, 75).

If Richard Fumerton's (2004) read of things is right, the appearance of internalist concessions at the second-order of Sosa's picture is specious. Fumerton's thinking here—drawing originally from a well-known point due to Laurence BonJour (1985)—is that there are two fundamentally different ways one might think about *how it is* that coherence is justification-conferring. On one way of thinking about things, the fact that a given belief coheres with other beliefs in one's doxastic system suffices to raise the epistemic status of the belief in question. Alternatively, and more demandingly, one might hold that the mere fact of a belief's cohering with other beliefs in one's system of beliefs does not alone confer justification upon (or otherwise raise the epistemic status of) the target belief *unless* the subject is *aware* of the fact that the belief coheres in this way. With this distinction in mind, Fumerton maintains that whatever boost to the epistemic status of a belief derives from the mere *fact* of its cohering with other beliefs would be one that is 'intellectually unsatisfying'. Though it's not clear from Sosa's proposal that he opts for any sort of further awareness requirement, or indeed, how such a requirement is something he could meet in a principled way.

Fumerton accordingly sees for Sosa a kind of dilemma, according to which:

> [...] coherence without access to coherence doesn't do the job of giving us the sort of justification that would satisfy an internalist. Without access requirements to coherence, however, it's not clear that we have given the internalist anything that would allow the internalist to view the internalism/externalism debate as a false dichotomy (2004, 81).

Fumerton is right that without access requirements to the kind of coherence that features for Sosa at the second-order, we likely won't satisfy an internalist, or at least, an accessibilist internalist. However, let's bear in mind the context of this criticism: virtue perspectivism does not aim at *internalism*, but at preserving some of the elements of a philosophical account of knowledge that the internalist values—something that pure 'thermometer' model reliablists (e.g., Armstrong 1973) are unable to do. A criticism according to which virtue perspectivism would not *satisfy* an internalist then misses the mark.

Secondly, there is an important sense in which a *kind* of access requirement really is satisfied in connection with the coherence one attains on virtue perspectivism at the second order. Indeed, when one transitions from animal knowledge that *p* to reflective knowledge that *p*, the kind of broad coherence that features at the second order furnishes the thinker with "a proper *awareness* of our competences through whose exercise we can gain our first-order knowledge" (Sosa 2017, 46)— and this is so even when, in transitioning from animal to reflective knowledge that *p*, one needn't be aware *that* p coheres in the relevant way when it does.

A principal value of internalism is that good epistemic standing involves not only the obtaining of certain epistemically good-making properties of our beliefs, but that we should be aware of their obtaining. Broad coherence at the second-order helps to provide a thinker with such awareness of one's good standing at the first order. This is accordingly a feature of virtue perspectivism that does justice to a key value of internalism in epistemology. And it needn't require one to have access to facts about a given belief's cohering when it does in order to do justice in this way.

A third line of response to Fumerton's critique requires us to return with a more critical eye to his distinction (originally due to Bonjour) between two kinds of coherence, one that is especially internalist friendly in that it involves not only coherence but awareness *of* coherence, and the other which is not and does not. Might we have positive reason, when giving an account of knowledge, to part ways with the kind of specifically internalist thinking about coherence that Fumerton rightly suggests virtue perspectivism isn't in a position to countenance? I think there is. Even more, one such argument to this effect can be extracted from one of Sosa's (1985) early papers on the value of coherence, one that pre-dates his bi-level epistemology.[16] One of Sosa's key insights in this early paper is that epistemic value of coherence is itself plausibly explained in terms of the value of reliability understood along externalist lines. In order to illuminate this idea, a thought experiment is presented, one involving a kind of 'random' world:

> Let us suppose [...] that beyond the causal regularity required at the mind/world interface there is no systematic and orderly depth either in the mind or in the world. Beyond a certain elementary level, in such a world coherent unity adds nothing to the likelihood of having the truth, and random scatter in our body of beliefs seems no less likely to get it right. How plausible is it to insist even for such a world that knowledge even of the mind/world interface is aided by the most elaborate possible webs both worldward and mindward? Surely it is very little plausible to suppose that such artificial and wholly false webs add anything at all to one's knowledge of what is there knowable. This suggests that coherence has derivative and not fundamental status as a source of cognitive justification. It justifies in our world, or so we believe, in virtue of its reliability as a source of truth (1985, 20).

The above thought experiment is meant to function as a *reductio* against the thought that coherence confers justification upon a given belief in a way that is *not* derivative upon its reliability. And this reductio, in our present context, motivates a kind of dilemma for Fumerton. The first horn of the dilemma is that, if the epistemic value of coherence is itself plausibly explained in terms of the value of reliability understood along externalist lines, then it's no problem for virtue perspectivism that "coherence without access to coherence doesn't do the job of giving us the sort of justification that would satisfy an internalist" (Fumerton 2004, 81). And this is because, put simply, the value of reliable coherence is more fundamental than the value of coherence with access to coherence. The other horn of the dilemma for the critic of virtue perspectivism is to deny that epistemic value of coherence is itself plausibly explained in terms of the value of reliability understood along externalist lines, but then to account for how the kind of coherence one might attain in Sosa's random world adds value to the knowledge one is able to acquire at the mind/world interface.

---

[16] See Sosa (1991) and, subsequently, Sosa (2007) and Sosa (2015).

8.8   Concluding Remarks

Virtue perspectivism, much like rationalist perspectivism, is an account of the nature of knowledge that emerges in direct response to questions about the very *possibility* of knowledge. Sosa's virtue perspectivism has changed in some of its peripheral details over the years, but the basic structure of the proposal, and in particular its anti-sceptical strategy, has remained the same: virtue perspectivism allows the use of our basic foundational faculties in attaining not only first-order animal knowledge without first knowing these faculties to be reliable, but also what Sosa calls a second-order assuring perspective, one whereby we can appreciate those first-order faculties *as* reliable and in doing so place our first-order knowledge in a competent second-order perspective.

Is this *problematically* circular? And, relatedly, is it even possible to vindicate the circularity that does seem to feature in the proposal as benign while at the same time doing justice to internalist intuitions in any meaningful sense? Can this even be done by a foundationalist proposal that builds its entire edifice not on the direct apprehension of anything that is 'given' in experience but rather by beliefs appropriately caused? Engaging seriously with such questions leads us, almost immediately, to the most fundamental questions in epistemological theory involving circularity, scepticism, doubt, and assurance. This essay has attempted to navigate at least some of these issues, with a focus on what I think are three especially rich criticisms of virtue perspectivism raised by Stroud, Reed, and Fumerton respectively. The conclusion reached is that virtue perspectivism can ultimately withstand these criticisms. Showing why this is the case involves (among other things) a careful engagement not only with the general dispute between internalists and externalists, but also with questions about what a philosophical theory of knowledge should be expected to do.

Bibliography

Armstrong, D. M. (1973). *Belief, truth and knowledge.* Cambridge: Cambridge University Press.

BonJour, L. (1985). *The structure of empirical knowledge.* Cambridge: Cambridge University Press.

Carter, J. A. (2016). *Metaepistemology and relativism.* London: Palgrave Macmillan.

Carter, J. A., & Littlejohn, C. (forthcoming). *This is epistemology.* Wiley-Blackwell.

Fumerton, R. (2004). Achieving epistemic ascent. In Greco, J (ed.), *Ernest sosa and his critics* (pp. 72–85).

Goldman, A. I. (1999). *Knowledge in a social world.* Oxford: Clarendon Press.

Greco, J. (2011). Epistemic circularity: Vicious, virtuous and benign. *International Journal for the Study of Skepticism, 1*(2), 105–112.

Hasan, A., & Fumerton R. (2017). Foundationalist theories of epistemic justification. In Edward N. Zalta (ed.), *The Stanford encyclopedia of philosophy*, Spring 2017. https://plato.stanford.edu/archives/spr2017/entries/justep-foundational/; Metaphysics Research Lab, Stanford University.

Moore, G. E. (1959). Four forms of skepticism. In *Philosophical papers* (pp. 193–223). New York: Macmillan.

Palermos, S. O. (2018). Epistemic presentism. *Philosophical Psychology, 31*(3), 458–478.

Reed, B. (2002). How to think about Fallibilism. *Philosophical Studies, 107*(2), 143–157.

Reed, B. (2012). Knowledge, doubt, and circularity. *Synthese, 188*(2), 273–287.

Rorty, R. (1979). *Philosophy and the Mirror of nature*. Princeton: Princeton University Press.

Sellars, W. (1956). Empiricism and the philosophy of mind. *Minnesota Studies in the Philosophy of Science, 1*(19), 253–329.

Sosa, E. (1985). The coherence of virtue and the virtue of coherence. *Synthese, 64*(1), 3–28.

Sosa, E. (1991). *Knowledge in perspective: Selected essays in epistemology*. Cambridge: Cambridge University Press.

Sosa, E. (1997a). Mythology of the given. *History of Philosophy Quarterly, 14*(3), 275–286.

Sosa, E. (1997b). Reflective knowledge in the best circles. *The Journal of Philosophy, 94*(8), 410–430.

Sosa, E. (1999). How to defeat opposition to Moore. *Noûs, 33*(13), 141–153.

Sosa, E. (2007). *A virtue epistemology: Apt belief and reflective knowledge* (Vol. 1). Oxford: Oxford University Press.

Sosa, E. (2009). *Reflective knowledge: Apt belief and reflective knowledge* (Vol. 2). Oxford: Oxford University Press.

Sosa, E. (2015). *Judgment and agency*. Oxford: Oxford University Press.

Sosa, E. (2017). *Epistemology*. Princeton: Princeton University Press.

Stroud, B. (2004). Perceptual knowledge and epistemological satisfaction. In J. Greco (Ed.), *Ernest Sosa and his critics* (pp. 165–173). Malden: Blackwell.

Chapter 9
Knowledge from a Human Point of View

Barry Stroud

Abstract Everything that is known by human beings is known from a human point of view. There is no other point of view from which human beings can know anything. Is there something distinctively "perspectival" about human knowledge or the study of human knowledge? Explaining how such-and-such has come to be known by human beings involves explaining how those who know it came to get things right. Those who explain that knowledge are thereby committed to the truth of what is said to be known. Can we explain, from a human point of view, how it has come to be known that such-and-such is so in the world we live in? Or can we explain, from a human point of view, only how it has come to be known that such-and-such is so from a human point of view in the world we live in?

Keywords Knowledge · Perspectivism · Scepticism · Truth

I am intrigued by the phrase "knowledge from a human point of view". It raises delicate questions. Human beings have known many things about the world for a long time. And we continue to learn more and more every day. And whatever we human beings come to know is of course known from a human point of view. There is no other point of view from which human beings could know anything. So one way to understand the phrase "knowledge from a human point of view" is to take it simply to refer to human knowledge: everything human beings know. That amounts by now to a huge and truly impressive body of knowledge. Of course, that body of knowledge constantly changes, as new things are learned and others are abandoned as not true and so never known. We can speak more cautiously of what is known by human beings at a certain time, or during a certain period. And of course that can change too.

But the phrase "knowledge from a human point of view" also speaks of a way of knowing things: from "the human point of view" through which the knowledge is gained. That "point of view" is obviously not simply a position in space and time:

B. Stroud (✉)
Philosophy Department, University of California, Berkeley, CA, USA
e-mail: barrys@berkeley.edu

A. Crețu, M. Massimi (eds.), *Knowledge from a Human Point of View*,
Synthese Library 416, https://doi.org/10.1007/978-3-030-27041-4_9

the total region of the universe occupied at some time by human beings, for instance. "Knowledge from a human point of view" presumably means knowledge gained by human means: through the exercise of distinctively human sensory and intellectual capacities. In asking how human beings know things by those means we are asking in effect how, given what human beings are like, and what the world they live in is like, human beings have come to know the things they know. The question is completely general, not only about institutionally-organized knowledge in the form of sciences. Science is part of it, of course, but the question is how human beings, with their needs and desires, their natural talents, common sense, rituals and lore, languages, interests, traditions, institutions, and practices come to know all the things they know. How has all the knowledge we think there is in the world come to be?

We know at least that it has all been acquired "from a human point of view". And, being human, we ask the question ourselves, unavoidably, from "a human point of view". Since human knowledge is what is in question, part of the knowledge each of us is interested in is our own knowledge. Each of us is asking, "from a human point of view", how each of us knows the things we know "from a human point of view". Can we really get a satisfactory understanding of human knowledge in that way? We can seem to be presented with a puzzle because each of us is at the same time both the subject and the object of our investigation. It is we, as agents, who want to understand how certain inhabitants of the world – we human beings – know the things we know about the world "from a human, viz. our, point of view".

Is this really a special difficulty? What exactly do we want to understand in this way? Do we expect to understand how human beings know, admittedly from a human point of view, what is so in the world they live in? Or is the most we can expect to understand only how human beings can know, from a human point of view, what is so from a human point of view in the world they live in? These sound like different goals, and different possible achievements. Which do we seek? Which do we expect? Which is better? Given that whatever we know we know from a human point of view, wouldn't we be left in a less satisfying, more restricted position if the most we could understand was only how human beings can know what is so from a human point of view in the world they live in, rather than understanding how human beings can know what is so in the world they live in?

On reflection, we might ask whether this apparently more restricted possibility even makes sense. What does it mean to say that something or other is so (or not) "from a certain point of view"? Not just that something is *believed* to be so or *known* to be so from a certain point of view, but that something *is so* (or not) from a certain point of view. Do we really understand how or what that could be? This is one delicate question raised by the phrase with which we began: does speaking of "knowledge from a human point of view" add any special dimension or difficulty to the problem of understanding human knowledge?

I wonder whether a worry along these lines might be part of what lies behind the appeal of "perspectivism", sometimes called "perspectivalism". I cannot say I am sure about exactly what that view is, or what it says, but I take perspectivism to be, very roughly, the idea that in investigating ourselves and our knowledge the most we are in a position to understand and explain – perhaps the most there is to understand – is our own attitudes and point of view on ourselves and the world: our *taking* ourselves to know things about the world, or *regarding* ourselves as knowers of the world we take ourselves to know. I will come back later to the question of perspectivism.

It might look as if how human beings come to know the things they know about the world they live in is a pretty straightforward question about how certain things happen in the world, or how one part of the world affects another. On the one hand there is the way things are in the world in all their purely non-human, impersonal aspects. And on the other, there is the way human beings are, with their distinctive capacities, talents, and traditions. That human beings have those distinctive characteristics is of course just as much a fact of the world as facts not involving human beings. And human beings with those characteristics come to know things about the world they live in. So it looks as if explaining the presence of human knowledge in the world would be a matter of explaining how human beings who exercise their natural capacities and practices in interacting with the surrounding world come to know how things are in the world they interact with. On this view, human knowledge acquired from a human point of view would be intelligible as a natural phenomenon in the world we live in.

I think there is something right, or at least promising, in this idea, but as it stands I think it cannot give us an explanation of human knowledge of the kind we seek. What is promising is the possibility of explaining in historical or developmental terms how human beings have come to believe all the things they believe. There must be some explanations of how all that happened, after all, whether we can actually explain it or not. But explaining how people have come to believe the things they believe, even if the explanation is correct as far as it goes, is not the same as explaining how they know the things they know. Something's being believed, even by many people in many different circumstances for a long time, is not the same as its being known.

There is something distinctive, and apparently more demanding, in understanding knowledge —as I think the history of philosophy amply illustrates. One fundamental difference is that knowledge implies truth; if something is known, it is true. That does not hold for belief. If different bodies of belief cannot all be true together, then not all of them are bodies of knowledge. And whether a body of beliefs does amount to knowledge or not cannot be determined simply by explaining how the beliefs in question came to be accepted. So historical or developmental explanation of the origin of beliefs is not enough to explain human knowledge.

The fact that what is known must be true explains why human knowledge is cumulative. To arrive at something not previously known by steps known to be reliable from something already known, or by finding it explicable only on the assumption of things already known, is to come to know something new. It is to add to the body of human knowledge, not simply to one's body of beliefs. Human knowledge grows because it is built on what is already known and so true. We are getting to know more and more about what is so every day, not just getting more beliefs. Of course, there is no guarantee of success. If we do know things, it is because human beings are successful in coming to know things. That is what looking at "knowledge from a human point of view" is meant to help us understand.

The fact that what is known must be true also helps explain what is a striking fact about human knowledge. Most of what comes to be common knowledge among human beings is learned from other people's knowing it, not by each person's reaching the same conclusion independently. This is important not only for the development of human knowledge, but for human life. There is just too much for each of us to know to do it all on our own; we simply couldn't get by without learning much of it from its being known by others and being available to us only in that way.

If what is known must be true, then if we come to believe that a certain person knows such-and-such, we too must acknowledge that what that person knows is true. We cannot stand apart and remain non-committal on the question of the truth of what we grant that other person knows. For the same reason, if we take seriously the idea of investigating human knowledge, and want to explain, even "from a human point of view", how human beings know certain things, we must agree that they do know those things, and therefore that the things they know are true. Of course, we might find that others do not really know certain things we thought they know, or even that they are not true. We would not then regard it as knowledge, not even "knowledge from a human point of view". But if we do take certain people to know certain things, and we ourselves are committed to the truth of what we take them to know, then to *explain* their knowledge we must explain their *knowing* what they know. That involves explaining not simply their believing it, or even their believing that they know it. It requires explaining their *getting it right*. That is what is distinctive of explaining knowledge. It is more demanding than explaining belief. It commits you to the truth of what you regard as known. It is more demanding even than explaining true belief. It commits you to explaining the knower's success in getting it right.

If I believe that a certain other person knows a certain thing, I believe that what that person knows is true. So I believe it and am prepared to act on it on the basis of my attributing knowledge of it to her. That does not yet mean that I myself know what I think that other person knows. It is not that easy to get to know something. We can think we know something when we don't: either it is not true, or we don't know it. If I were wrong in thinking that the other person knows what I think she knows, I would not know it either. But if I nonetheless believe it, and it is in fact true, then what I believe as a result of relying on her knowing it is true. In accepting it and acting on it, I have a true belief. That true belief might become widely enough shared by others even to be part of what is called "common knowledge". It would be true, and widely believed, even if it turns out that nobody in fact knows it. It would then not be part of human knowledge.

We accept something we believe to be known by others because we take ourselves to know or have good reason to believe that those others do know it. In accepting it we believe their grounds or reasons for claiming to know it are sound and adequate for knowing. That is how we learn things from scientific and historical books, for instance. We believe that the authors of the books know what they are talking about and know that what they say is true. In accepting what the books say we implicitly make positive evaluations of the competence and judgement of the authors, and of the correctness of what they say in the case in question. Making sound judgements of the competence of others is essential to our recognizing them as knowing things, and so learning from them. Judgement of that kind is part of our very conception of human knowledge. Not our conception of what is known by those who know, but our conception of what it is for somebody to know something.

The kinds of capacities, judgements, and abilities we credit others with in ascribing knowledge to them are the same qualities we ourselves possess and exercise in knowing the things we know. We all gradually acquire such capacities by learning to speak and understand the things we and other people say. To know something we must be capable of understanding and endorsing thoughts that express what we know. Since what is known must be true, they must be thoughts of something or other's being so: predicational thoughts with a truth value. Endorsing such thoughts is a matter of judging or putting it forward that what they express is true. We must be, or eventually become, masters of the procedures and discriminations necessary for having such thoughts and judging some of them to be true, if we are to have even as much as a capacity for knowing the kinds of things human beings come to know.

This is a gradual learning process; we do not go from blankness to a full repertoire of understanding and knowledge overnight. The learning takes place in the very circumstances we are learning to think about and make judgements about. To show that we have that capacity we must show repeatedly in practice that we are good at recognizing when the things we think we understand are in fact true, or, as the case might be, not true. It is easiest to do this, to begin with, in simple observational circumstances, when the objects we claim to know about are right before us. By getting more and more of the things we think in those circumstances right we are gaining more and more of the conceptual resources needed for thinking and so knowing this or that to be so in the world. Even in the process of acquiring that capacity for knowledge we come to know things about the world. As our expertise as knowers grows, and more and more of the judgements we make are correct, our knowledge of the world is growing accordingly. We become masters of speech and thought, and acquire an elaborate, sensitive capacity for knowing things of many different kinds. But, of course, through it all we remain fallible human beings.

We do not do all this on our own, in isolation. It is only because there are common languages into which all of us are socialized that there is even such a thing as our meaning one thing rather than another by the sounds we utter or being understood by others to be saying such-and-such. A social practice is needed to provide us with a shared means of expressing ourselves and understanding one another. It is in those terms that we express what we all come to know about the world around us. We find ourselves, and must find ourselves, in a "human point of view" that is widely shared. Human knowledge, and the human thought and understanding that is required for it, is a social achievement. I think it is only from some such communal position — shall we say "from a human point of view" — that human beings come to understand one another and share the thoughts and judgements they must master even to be capable of knowing the things they all know about the world around them.

I think this conception of "a human point of view" from which to survey human knowledge would not be congenial to perspectivism as I understand it. Anyone occupying such a "point of view" already knows, and must know, many things about what is so in the world. It is not a "point of view" one can take up or occupy independently of, or somehow "from outside", one's knowing many things about the world. It is not non-committal on the question of whether there is human knowledge, or on the question of how things actually are in the world. To occupy this kind of "human point of view" is to be fully engaged in the community of human knowers and so to be committed to the world's being the ways it is widely known to be. That is to believe the world to be a certain way – all those ways one takes the world to be known to be. So human beings investigating human knowledge from "a human point of view" like this would find the world they investigate to be very much as they know it to be, and they would find it to be populated by other human beings who for the most part know the world to be just the ways they themselves know it to be.

This is certainly one kind of "human point of view" from which we can be presented with the problem of knowledge with which we began: how do human beings come to know the things they know? What, then, is the attitude or "human point of view" a perspectivist takes towards the human knowledge he or she would account for? It seems it could not be the straightforward acceptance of human beings' knowing the things they know, and so accepting the truth of the things they know, as I have just described it. There is nothing distinctively perspectival in that, or in the conception of human knowledge that is involved in it. Does the perspectivist focus, rather, on something else: not directly on human knowledge as such, but on human beings' taking certain attitudes towards themselves and the world: their *regarding* themselves as enquirers or knowers, their *taking* themselves to know things about the world, and so their *taking* the world to be a certain way? We human beings do have such attitudes towards ourselves and others, and we can find those attitudes to be present when we consider human knowledge "from a human point of view". Are those the attitudes perspectivism concentrates on?

There has been a long tradition in philosophy of finding the idea of human knowledge in itself suspicious, problematic, maybe even an impossibility. What is knowledge, anyway? Can it be defined? Can we ever really know anything? Can we even understand how we know something? Since at least the time of Plato the distinctive demands of knowledge have seemed to present a special problem. Sceptics in antiquity thought they could get along in life —or would even be better off — without seeking knowledge at all. Simply "going along with appearances" was a path to tranquillity.

In later centuries, more was thought to be needed for serious science, and beliefs, hypotheses, and theories became central to the philosophical understanding of human knowledge. Justification, confirmation, or reasons for accepting beliefs or theories, was what mattered. Justified belief was the goal, especially with consensus about the strength of the reasons. It was widely assumed that all that had to be added to a well-confirmed belief for it to be knowledge was that the justified belief be true. Alas, that has turned out not to be so. The idea of knowledge apparently cannot be explained as simply a combination of certain states of mind and states of affairs, each in itself less than knowledge. What, then, does knowledge amount to? Can it be defined at all? What role does the idea of knowledge actually play in the constant human effort to find out how things are in the world?

The most extreme reaction to difficulties in the idea of knowledge was present almost from the beginning: the sceptical conclusion that nobody can ever really know anything about the world around us. I think that view of the human condition is, as a simple matter of fact, not correct. I think there is no question that we all know a great many things, and that it is possible, in general, to explain how we know them. What I find most interesting, fascinating, and challenging about philosophical scepticism is not the flat-footed question whether it is true, or false, but what it really is, how it works, or how it is supposed to work.

I think there is a lot to be learned about ourselves and our thought about ourselves and the world by looking carefully into the conditions we can see to be necessary even for us to think of a world at all, and to understand ourselves as competently exercising the concepts we need in order to perceive and believe things about it. We could not even entertain the possibility of human knowledge if we did not fulfil those conditions. So, rather than flatly denying philosophical scepticism, or ignoring it, I think it is more fruitful to try to discover whether we could actually fulfil all those necessary conditions, and so whether we could even be presented with what looks like a general challenge to our knowledge of the world, if we did not also in fact (perhaps unrecognized by us at the moment) also know many things about that world, or at least take ourselves to know them. That would not show directly that philosophical scepticism is not true, but it might change our attitudes towards it. We might even come to see how and why that sceptical conclusion, for all its apparent force, is something we simply could never consistently accept, given that we possess what it takes even to understand it. Philosophical scepticism might then take on a different interest for us. If we did come to see and feel something like that, we could be said to have discovered it by reflecting on human knowledge "from a human point of view".

The tradition of looking askance at the concept of knowledge has not always gone as far as denying the very possibility of knowledge. Many have questioned whether we even need such a puzzling notion to account for the obvious success human beings have had in coping with the world and with one another in the ways they have. Could it be that perspectivism perhaps expresses a certain sympathy with this tradition of doubt or suspicion about knowledge? I speculate here, but the idea of knowledge is so directly connected with the idea of truth, which is independent of human beings' holding the attitudes they do towards it, that perhaps perspectivism sees more promise in shifting the focus away from the idea of knowledge as such, and looking instead to other human attitudes or responses involved in explaining what we want to understand about the whole enterprise of what we call human knowledge. I will end with one thought about this.

If this is a way to understand the appeal of perspectivism, I would suggest a line of thought in response to it parallel to my proposal about philosophical scepticism. First would be to ask what we primarily want to understand about the acquisition and development of what we call human knowledge. Is it human acceptance —and rejection — of more and more theories or hypotheses that we think needs accounting for? Or is it the fact of theory change, or the competition among theories: how can we tell which is best? Or is what we want to account for the progressive accumulation of more and more of what we call human knowledge: the reliable growth of an expanding conception of the world and of human life? And whatever the goal, can we really understand what we most want to understand about the enterprise of human knowledge by thinking of those who investigate the world as exercising only the concepts needed for the less-committal epistemic attitudes and responses that perspectivism concentrates on, not a concept of knowledge that implies truth and so apparently resists perspectival treatment?

This is a question I recommend to aspiring perspectivists. As I think I have found with philosophical scepticism, I think it promises, at the very least, a deeper understanding of the puzzles that confront us.

Index

Printed in the United States
by Baker & Taylor Publisher Services